Lecture Notes of
the Unione Matematica Italiana

28

Unione
Matematica
Italiana

Bozhidar Velichkov

Regularity of the One-phase Free Boundaries

 Springer

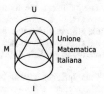

Unione
Matematica
Italiana

Bozhidar Velichkov
Department of Mathematics
University of Pisa
Pisa, Italy

This work was supported by Horizon 2020 Framework Programme (ERC project VAREG n.853404).

ISSN 1862-9113   ISSN 1862-9121   (electronic)
Lecture Notes ofthe Unione Matematica Italiana
ISBN 978-3-031-13237-7   ISBN 978-3-031-13238-4   (eBook)
https://doi.org/10.1007/978-3-031-13238-4

Mathematics Subject Classification: 35R35

This Springer imprint is published by the registered company Springer Nature Switzerland AG
The registered company address is: Gewerbestrasse 11, 6330 Cham, Switzerland

The Unione Matematica Italiana (UMI) has established a bi-annual prize, sponsored by Springer-Verlag, to honor an excellent, original monograph presenting the latest developments in an active research area of mathematics, to which the author made important contributions in the recent years.

The prize-winning monographs are published in this series. Details about the prize can be found at:

http://umi.dm.unibo.it/en/unione-matematica-italiana-prizes/book-prize-unione-matematica-italiana/

This book has been awarded the 2019 Book Prize of the Unione Matematica Italiana.

The members of the scientific committee of the 2019 prize were:

Piermarco Cannarsa (Presidente of the UMI)
University of Rome "Tor Vergata", Italy

Camillo de Lellis
IAS, Princeton, USA

Anna Maria Fino
University of Turin, Italy

Annamaria Montanari
University of Bologna, Italy

Valentino Tosatti
McGill University, Québec, Canada

Preface

This book is an introduction to the regularity theory for free boundary problems. The focus is on the one-phase Bernoulli problem, which is of particular interest as it deeply influenced the modern free boundary regularity theory and is still an object of intensive research. The exposition is organized around four main theorems, which are dedicated to the one-phase functional in its simplest form. Many of the methods and the techniques we present here are very recent and were developed in the context of different free boundary problems. We also give the detailed proofs of several classical results, which are based on some universal ideas and are recurrent in the free boundary, PDE, and the geometric regularity theories.

Pisa, Italy

Bozhidar Velichkov

Acknowledgment

The work of several authors deeply influenced the exposition, in particular Alt-Caffarelli [3], Alt-Caffarelli-Friedman (see [4] and Sect. 3.2), Briançon-Hayouni-Pierre ([7] and Sect. 3.2), Danielli-Petrosyan (see [18] and Sect. 3.3), Weiss (see [52] and Sect. 9), Federer (see [32] and Sect. 10), Garofalo-Lin (see [34] and Sect. 11.3), De Silva (see [23] and Sect. 8), Reifenberg (see [45] and Sect. 12.5), Bucur (see [8] and Sect. 5.2), and Briançon and Lamboley (see [5], [6] and Sect. 11.5).

Most of the present book consists of my personal notes, taken during the preparation of [41], [49], [42], [10], [29], [46] and [50]; I warmly thank my co-authors Luca Spolaor [49, 50], Dario Mazzoleni and Susanna Terracini [41, 42], Giuseppe Buttazzo [10], Max Engelstein [29], Emmanuel Russ [46], and BaptisteTrey [46, 50] for being part of this journey.

Part of these notes was presented in my lectures *"Regularity for free boundary and shape optimization problems"*; the course was held in Naples in 2019 and was part of the Indam Intensive Period *"Shape optimization, control and inverse problems for PDEs."* We warmly thank the organizers for this opportunity.

The author was supported by the European Research Council (ERC), under the European Union's Horizon 2020 research and innovation program, through the project ERC VAREG—*Variational approach to the regularity of the free boundaries* (grant agreement No. 853404).

Contents

Chapter 1
Introduction and Main Results

1.1 Free Boundary Problems: Classical and Variational Formulations

The *free boundary problems* are a special type of boundary value problems, in which the domain, where the PDE is solved, depends on the solution of the boundary value problem. A classical example of a free boundary problem is the Serrin problem:

> *Find a bounded open C^2−regular connected domain $\Omega \subset \mathbb{R}^d$*
>
> *and a function $u : \Omega \to \mathbb{R}$ such that :*
>
> $-\Delta u = 1 \quad \text{in} \quad \Omega, \qquad u = 0 \quad \text{and} \quad |\nabla u| = c \quad \text{on} \quad \partial\Omega.$

It is well-known (see [47]) that, up to translation, the unique solution of the Serrin problem is given by the couple (B, w_B), where B is the ball of radius $R = d$ (d is the dimension of the space) and $w_B : B \to \mathbb{R}$ is the function $w_B(x) = \frac{1}{2d}\left(R^2 - |x|^2\right)$.

More generally, if D is a smooth bounded open set in $\mathbb{R}^d$, then we can consider the following problem. Find a couple (Ω, u) such that:

- the domain Ω is contained in D
- while the function $u : \Omega \to \mathbb{R}$

  - solves a PDE in Ω, which in the example (1.1) below (as in the rest of these notes) is elliptic but, in general, can also involve a time variable:

$$\sum_{i,j=1}^{d} a_{ij}(x)\partial_{ij}u + \sum_{i=1}^{d} b_i(x)\partial_i u + c(x)u(x) = f(x) \quad \text{in} \quad \Omega; \qquad (1.1)$$

© The Author(s) 2023
B. Velichkov, *Regularity of the One-phase Free Boundaries*,
Lecture Notes of the Unione Matematica Italiana 28,
https://doi.org/10.1007/978-3-031-13238-4_1

– satisfies a boundary condition on the fixed boundary ∂D, that is,

$$F(x, u, \nabla u) = 0 \quad \text{on} \quad \partial D \cap \partial \Omega; \tag{1.2}$$

– satisfies an overdetermined boundary condition on the free boundary $\partial \Omega \cap D$

$$G(x, u, \nabla u) = 0 \quad \text{and} \quad H(x, u, \nabla u) = 0 \quad \text{on} \quad \partial \Omega \cap D, \tag{1.3}$$

where the functions $F, G, H : \mathbb{R}^{2d+1} \to \mathbb{R}$, as well as the elliptic operator and the right-hand side in (1.1), are given. The aim of the free boundary regularity theory is to describe the interaction between the free boundary $\partial \Omega$ and the solution u of the PDE. For instance, it is well-known that, the solutions of boundary value problems (with sufficiently smooth data) inherit the regularity of the boundary $\partial \Omega$, that is, if $\partial \Omega$ is $C^{1,\alpha}$, then $|\nabla u|$ is Hölder continuous up to the boundary (see [35]). Conversely, one can ask the opposite question. Suppose that u is a solution of the free boundary problem (1.1)–(1.3), where the overdetermined condition (1.3) on the free boundary is given by

$$u = 0 \quad \text{and} \quad |\nabla u|^2 = Q(x) \quad \text{on} \quad \partial \Omega \cap D,$$

for some Hölder continuous function Q. Is it true that $\partial \Omega$ is $C^{1,\alpha}$-regular? More generally, we can ask the following question:

> Is it possible to obtain information on the local structure of the free boundary, just from the fact that the overdetermined boundary value problem admits a solution?

Notice that, here we do not impose any a priori regularity on the domain Ω. For an extensive introduction to the free boundary problems, with numerous concrete examples and applications, we refer to the book [33], while a more advanced reading is [15].

A free boundary problem of particular relevance for the theory is the so-called *one-phase Bernoulli problem*, which was the object of numerous studies in the last 40 years; it also motivated the introduction of several new tools and the development of new regularity techniques. The problem is the following. We have given:

- a smooth bounded open set D in $\mathbb{R}^d$,
- a non-negative function $g : \partial D \to \mathbb{R}$,
- a positive constant Λ,

and we search for a couple (Ω, u), of a domain $\Omega \subset D$ and a function $u : \Omega \to \mathbb{R}$, such that:

$$\begin{cases} \Delta u = 0 & \text{in} \quad \Omega, \\ u = g & \text{on} \quad \partial \Omega \cap \partial D, \\ u = 0 \quad \text{and} \quad |\nabla u| = \sqrt{\Lambda} & \text{on} \quad \partial \Omega \cap D. \end{cases} \tag{1.4}$$

Fig. 1.1 A minimizer u and its free boundary; for simplicity we take $D = B_1$

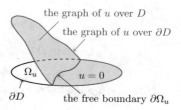

the graph of u over D

the graph of u over ∂D

Ω_u

$u = 0$

∂D

the free boundary $\partial \Omega_u$

We notice that a solution should depend both on the ambient domain D and the boundary value g. Thus, we cannot hope to find explicitly the domain Ω and the function u, except in some very special cases. In fact, even the existence of a couple (Ω, u) solving (1.4) is a non-trivial question. One way to solve the existence issue is to consider the variational problem, which consists in minimizing the functional

$$u \mapsto \mathcal{F}_\Lambda(u, D) = \int_D |\nabla u|^2 \, dx + \Lambda |\{u > 0\} \cap D|,$$

among all functions $u : D \to \mathbb{R}$ such that

$$u \in H^1(D) \qquad \text{and} \qquad u = g \quad \text{on} \quad \partial D.$$

A solution to (1.4) can be obtained in the following way (see Fig. 1.1). To any minimizer $u : D \to \mathbb{R}$, we associate the domain

$$\Omega_u := \{u > 0\},$$

and the free boundary $\partial \Omega_u \cap D$. Then, at least formally, one can show that the couple (Ω_u, u) is a solution to the free boundary problem (1.4).

- First, notice that the conditions

$$u = 0 \quad \text{on} \quad \partial \Omega_u \cap D,$$

$$u = g \quad \text{on} \quad \partial \Omega_u \cap \partial D,$$

are fulfilled by construction.
- In order to show that u is harmonic in Ω_u, we suppose that Ω_u is open and that u is continuous. Let $\varphi \in C_c^\infty(\Omega_u)$ be a smooth function of compact support in Ω_u. Then, for any $t \in \mathbb{R}$ sufficiently close to zero, we have

$$\{u + t\varphi > 0\} = \{u > 0\},$$

and so,

$$\mathcal{F}_\Lambda(u + t\varphi, D) = \mathcal{F}_\Lambda(u, D) + \int_{\Omega_u} \left(|\nabla(u + t\varphi)|^2 - |\nabla u|^2 \right) dx.$$

Now, the minimality of u gives that

$$2 \int_{\Omega_u} \nabla u \cdot \nabla \varphi \, dx = \frac{\partial}{\partial t}\Big|_{t=0} \mathcal{F}_\Lambda(u + t\varphi, D) = 0.$$

Integrating by parts and using the fact that φ is arbitrary, we get that

$$\Delta u = 0 \quad \text{in} \quad \Omega_u.$$

• Finally, for what concerns the overdetermined condition on the free boundary, we proceed as follows. For any compactly supported smooth vector field $\xi :$ $D \to \mathbb{R}^d$ and any (small) $t > 0$, we consider the diffeomorphism $\Psi_t(x) = x + t\xi(x)$ and the test function $u_t = u \circ \Psi_t^{-1}$. Then, by the optimality of u, we obtain

$$0 = \frac{\partial}{\partial t}\Big|_{t=0} \mathcal{F}_\Lambda(u_t, D).$$

On the other hand, the derivative on the right-hand side can be computed explicitly (see Lemma 9.5). Precisely, if we assume that u and $\partial\Omega_u$ are smooth enough, we have

$$\frac{\partial}{\partial t}\Big|_{t=0} \mathcal{F}_\Lambda(u_t, D) = \int_{\partial\Omega_u} \left(-|\nabla u|^2 + \Lambda \right) \xi \cdot v \, d\mathcal{H}^{d-1},$$

where v is the exterior normal to $\partial\Omega_u$. Since ξ is arbitrary, we get that

$$|\nabla u| = \sqrt{\Lambda} \quad \text{on} \quad \partial\Omega_u \cap D.$$

In conclusion, by minimizing the function $\mathcal{F}_\Lambda$, we obtain at once the function u and the domain Ω solving (1.4). The function u is a minimizer of $\mathcal{F}_\Lambda$ and the set Ω is defined as $\Omega = \Omega_u = \{u > 0\}$. The equation in Ω_u and the overdetermined condition on the free boundary $\partial\Omega_u \cap D$ are in fact the Euler-Lagrange equations associated to the functional. Thus, instead of studying directly the free boundary problem (1.4), in these notes, we will restrict our attention to minimizers of $\mathcal{F}_\Lambda$. In order to fix the terminology and the notations in this section, and also for the rest of these notes, we give the following definition.

Definition 1.1 (Minimizers of $\mathcal{F}_\Lambda$) Let D be a bounded open set in $\mathbb{R}^d$. We say that the function $u : D \to \mathbb{R}$ is a minimizer of $\mathcal{F}_\Lambda$ in D, if $u \in H^1(D)$, $u \geq 0$ in D and

$$\mathcal{F}_\Lambda(u, D) \leq \mathcal{F}_\Lambda(v, D) \qquad \text{for every} \qquad v \in H^1(D) \quad \text{such that} \quad u - v \in H_0^1(D).$$

1.2   Regularity of the Free Boundary

These notes are an introduction to the free boundary regularity theory; the aim is to describe the local structure of the free boundary $\partial \Omega_u$ (which is a geometric object) just by using the fact that u minimizes the functional $\mathcal{F}_\Lambda$ and solves an overdetermined boundary value problem (that is, with techniques coming from Calculus of Variations and PDEs). In fact, the free boundary regularity theory stands on the crossroad of Calculus of Variations, PDEs and Geometric Analysis, and is characterized by the interaction between geometric and analytic objects, which is precisely what makes it so fascinating (and hard) field of Analysis.

Our aim in these notes is to prove a first theorem on the local structure of the free boundary. In particular, just by using the fact that u is a minimizer of the functional $\mathcal{F}_\Lambda$, we will prove the following facts:

- $u : D \to \mathbb{R}$ is (locally) Lipschitz continuous;
- the set $\Omega_u := \{u > 0\}$ is open and the free boundary $\partial \Omega_u \cap D$ can be decomposed as the disjoint union of a regular part, $Reg(\partial \Omega_u)$, and a singular part, $Sing(\partial \Omega_u)$,

$$\partial \Omega_u \cap D = Reg(\partial \Omega_u) \cup Sing(\partial \Omega_u),$$

  for instance, as on Fig. 1.2;
- the regular part $Reg(\partial \Omega_u)$ is a $C^{1,\alpha}$-smooth manifold of dimension $(d - 1)$;
- the singular part $Sing(\partial \Omega_u)$ is a closed subset of $\partial \Omega_u \cap D$ and its Hausdorff dimension is at most $d - 3$ (at the moment, the best known estimate for the Hausdorff dimension of the singular set is $d - 5$).

The overall approach and many of the tools that we will present are universal and have counterparts in other fields, for instance, in the regularity of area-minimizing

Fig. 1.2 A picture of a free boundary $\partial \Omega_u$ with regular and singular points

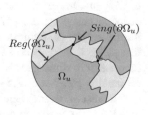

currents, in free discontinuity problems and harmonic maps. In fact, there are several points which are common for the regularity theory in all these (and many other) variational problems:

— the local behavior of the solution is determined through the analysis of the so-called blow-up sequences and blow-up limits;
— the points of the free boundary are labelled *regular* or *singular* according to the structure of the so called *blow-up limits* at each point; this provides a decomposition of the free boundary into a regular part and a singular part;
— at *regular* points, the regularity of the free boundary, which might be expressed in geometric (Theorem 7.4) or energetic (Theorem 12.1 and Lemma 12.14) terms, improves along the blow-up sequences;
— the set of *singular* points can become bigger when the dimension of the ambient space is higher; the measure and the dimension of the singular set can be estimated through the so-called dimension reduction principle, which uses the fact the blow-up limit are homogeneous functions; the homogeneity of the blow-up limits can be obtained through a monotonicity formula.

We will prove four main theorems.

In Theorem 1.2 (Sect. 1.3) we prove a regularity result for minimizers of $\mathcal{F}_\Lambda$. We will obtain the $C^{1,\alpha}$ regularity of the *regular part* of the free boundary through an improvement-of-flatness approach, while we will only give a weak estimate on the measure of the *singular set*. The proof of this theorem is carried out through Chaps. 2–8.

In Theorem 1.4 (Sect. 1.4) we give an estimate on the dimension of the set of *singular points*. We will use the Weiss monotonicity formula to obtain the homogeneity of the blow-up limits and the Federer dimension reduction principle to estimate the dimension of the singular set. The proof of this theorem is contained in Chaps. 9 and 10.

In Theorem 1.9 (Sect. 1.5) we prove a regularity theorem for functions u minimizing $\mathcal{F}_0$ under the additional measure constraint $|\Omega_u| = m$. In this case, we show that there is a Lagrange multiplier Λ such that u is a *critical point* for the functional $\mathcal{F}_\Lambda$. In this case, the regularity of the free boundary is a more delicate issue and the Theorems 1.2 and 1.4 cannot be applied directly. The proof requires the Chaps. 2–10, and also the specific analysis from Chap. 11.

Theorem 1.10 (Sect. 1.6) is dedicated to the epiperimetric inequality (Theorem 12.1) approach to the regularity of the free boundary, which was introduced in [49]. In particular, we give another proof of the fact that, if u is a local minimizer of $\mathcal{F}_\Lambda$ in dimension two, then the (entire) free boundary is $C^{1,\alpha}$ regular. The proof is based on the epiperimetric inequality from Sect. 12, which replaces the improvement of flatness argument from Chap. 7, but we still use results from Chaps. 2, 3, 4, 6, 8 and 9. Finally, we notice that the fact that an epiperimetric inequality in dimension d implies the regularity of the free boundary holds in any dimension (see Sect. 12.5).

The rest of the introduction is organized as follows. Each of the Sects. 1.3, 1.4, 1.5 and 1.6 is dedicated to one of the main theorems 1.2, 1.4, 1.9 and 1.10. Finally, in

Sect. 1.7, we briefly discuss some of the results, obtained or just reported in these notes, which might also be of interest for specialists in the field.

1.3   The Regularity Theorem of Alt and Caffarelli

Alt and Caffarelli pioneered the study of the one-phase free boundaries in [3], where they proved the following theorem.

Theorem 1.2 (Alt-Caffarelli) *Let D be a bounded open set in $\mathbb{R}^d$ and $u \in H^1(D)$ be a non-negative minimizer of $\mathcal{F}_\Lambda$ in D. Then u is locally Lipschitz continuous in D, the set $\Omega_u = \{u > 0\}$ is open and the free boundary can be decomposed as:*

$$\partial \Omega_u \cap D = Reg(\partial \Omega_u) \cup Sing(\partial \Omega_u),$$

where $Reg(\partial \Omega_u)$ and $Sing(\partial \Omega_u)$ are disjoint sets such that:

(i) $Reg(\partial \Omega_u)$ is a $C^{1,\alpha}$-regular $(d-1)$-dimensional surface in D, for some $\alpha > 0$;
(ii) $Sing(\partial \Omega_u)$ is a closed set of zero $(d-1)$-dimensional Hausdorff measure.

In these notes we will give a proof of this result, which is different from the original one (see [3]) and is based on recent methods developed in several different contexts: for instance, the two-phase problem [4, 50], almost-minimizers for the one-phase problem [19, 50], the one-phase problem for singular operators [18], the vectorial Bernoulli problems [41, 42], shape optimization problems [9, 46]. We will also use tools, which were developed after [3] as, for instance, viscosity solutions [12], [13], [14], [23], [26] and [15], monotonicity formula [52] and epiperimetric inequalities [29, 49].

In order to make these notes easier to read, we give the sketch of the proof in the introduction; for the technical details and generalizations, we refer to the results from the forthcoming chapters.

Proof In the proof of Theorem 1.2 we will use only results from Chaps. 2–8.

Section 2 is dedicated to the existence of minimizers and also to several explicit examples and preliminary results that will be useful in the forthcoming sections. The existence of minimizers for fixed boundary datum on ∂D is obtained in Proposition 2.1. In Lemma 2.6 and Lemma 2.7 we give two different proofs of the fact that the minimizers of $\mathcal{F}_\Lambda$ are subharmonic functions. This result has several important applications. First of all, when we study the local behavior of u and of the free boundary $\partial \Omega_u$, we may assume a priori that the function u is bounded. Moreover, as for a subharmonic function, the limit

$$\lim_{r \to 0} \fint_{B_r(x_0)} u(x)\, dx$$

exists at every point $x_0 \in \mathbb{R}$, we may also assume that u (which is a priori a Sobolev function, so defined as a class of equivalence of Lebesgue measurable functions) is defined pointwise *everywhere* in D. In particular, we will always work with the precise representative of u, defined by

$$u(x_0) = \lim_{r \to 0} \fint_{B_r(x_0)} u(x)\, dx \qquad \text{for every} \qquad x_0 \in D.$$

In particular, the set $\Omega_u = \{u > 0\}$ and its topological boundary $\partial \Omega_u$ are also well-defined (for all these results, we refer to Proposition 2.1). Moreover, in Lemma 2.9, we prove that the topological boundary coincides with the measure-theoretic one in the following sense:

$$\partial \Omega_u \cap D = \Big\{ x \in D \ : \ |B_r(x) \cap \Omega_u| > 0 \quad \text{and} \quad |B_r(x) \cap \{u = 0\}| > 0, \ \forall r > 0 \Big\}.$$

In Chap. 3 we prove that the function $u \ : \ D \ \to \ \mathbb{R}$ is locally Lipschitz continuous in D (Theorem 3.1). The main result of this section is more general (see Theorem 3.2) as for the Lipschitz continuity of u we only use that minimality of the function with respect to outwards perturbations.

We give three different proofs of the local Lipschitz continuity, inspired by three different methods, which were developed in the contexts of different free boundary problems. In Sect. 3.1, we report the original proof of Alt and Caffarelli; in Sect. 3.2, we give a proof which is inspired from the two-phase problem of Alt-Caffarelli-Friedman and already proved to be useful in several different contexts, for instance, for vectorial problems (see [9]) and for operators with drift (see [46]); in Sect. 3.3, we present the proof of Danielli and Petrosyan, which was originally introduced to deal with free boundary problems involving the p-Laplacian (see [18]); each of these subsections can be read independently.

As a consequence of the Lipschitz continuity, we obtain that the set Ω_u is open. Now, from the fact that u minimizes $\mathcal{F}_\Lambda$, we deduce that u is harmonic on Ω_u:

$$\Delta u = 0 \quad \text{in} \quad \Omega_u \cap D.$$

In particular, u is C^∞ regular (and analytic) in Ω_u.

In Chap. 4 (see Lemma 4.4 and/or Lemma 4.5), we prove that u is non-degenerate at the free boundary, that is, there is a constant $\kappa > 0$ such that the following claim holds:

If $x_0 \in \overline{\Omega}_u \cap D$, then $\|u\|_{L^\infty(B_r(x_0))} \geq \kappa r$, for every $r > 0$ such that $B_r(x_0) \subset D$.

This means that the Lipschitz estimate from Chap. 3 is optimal at the free boundary. This is a technical result, which we will use several times throughout the proof of Theorem 1.2, for instance, in Chaps. 5, 6 and 8.

In Chap. 5 we use the Lipschitz continuity and the non-degeneracy of u to obtain several results on the measure-theoretic structure of the free boundary. We will use this information in Sect. 6.4 to prove that the singular set has zero $(d-1)$-Hausdorff measure. The main results of Chap. 5 are the following:

- In Sect. 5.1 (Lemma 5.1), we prove that there is a constant $c \in (0, 1)$ such that, for every $x_0 \in D$ and every radius r small enough,

$$c|B_r| \leq |\Omega_u \cap B_r(x_0)| \leq (1 - c)|B_r|.$$

  In particular, the free boundary cannot contain points of Lebesgue density 0 or 1.
- In Sect. 5.2 (see Proposition 5.3 and Corollary 5.4), we prove that the set Ω_u has locally finite perimeter in D. We will use this result in Sect. 6.4 in order to estimate the dimension of the singular set.
- In Sect. 5.3 (Proposition 5.7), we prove that the free boundary $\partial \Omega_u \cap D$ has locally finite $(d-1)$-dimensional Hausdorff measure, which is slightly more general result than the one from Corollary 5.4.

Section 6 is dedicated to the convergence of the blow-up sequences and the analysis of the blow-up limits; both being essential for determining the local structure of the free boundary. The notion of a blow-up is introduced in the beginning of Chap. 6 (see Definition 6.1). For convenience of the reader, we anticipate that

for every $x_0 \in \partial \Omega_u \cap D$ and every infinitesimal sequence $(r_n)_{n \geq 1}$,

the sequence of rescalings

$$u_{x_0, r_n}(x) := \frac{1}{r_n} u(x_0 + r_n x)$$

is called a blow-up sequence at x_0. The (local) Lipschitz continuity of $u : D \to \mathbb{R}$ implies that, up to a subsequence, u_{x_0, r_n} converges to a globally defined Lipschitz continuous function $u_0 : \mathbb{R}^d \to \mathbb{R}$. Any function u_0 obtained in this way is called a blow-up limit of u at x_0. Notice that the non-degeneracy of u implies that u_0 cannot be constantly zero. In Proposition 6.2 we prove that the blow-up limit u_0 is a global minimizer of $\mathcal{F}_\Lambda$ (see Sect. 6.1) and that the free boundaries $\partial \{u_{x_0, r_n} > 0\}$ converge to $\partial \{u_0 > 0\}$ locally in the Hausdorff distance (Sect. 6.2).

In Sect. 6.4, we decompose the free boundary into regular and singular parts (see Definition 6.10), $Reg(\partial \Omega_u)$ and $Sing(\partial \Omega_u) := (\partial \Omega_u \cap D) \setminus Reg(\partial \Omega_u)$. Precisely, we say that a point $x_0 \in \partial \Omega_u \cap D$ is regular, if there is a blow-up limit u_0, of u at x_0, of the form

$$u_0(x) = \sqrt{\Lambda} \, (x \cdot \nu)_+ \tag{1.5}$$

for some unit vector ν. We then prove (see Lemma 6.11) that the regular part $Reg(\partial \Omega_u)$ contains the reduced boundary $\partial^* \Omega_u \cap D$. This is a consequence to the following two facts: first, at points of the reduced boundary $x_0 \in \partial^* \Omega_u \cap D$, the support of the blow-up limits is precisely a half-space $\{x : x \cdot \nu > 0\}$; second, if u_0 is a global solution supported on a half-space, then it has the form (1.5). This implies that $\mathcal{H}^{d-1}\big(Sing(\partial \Omega_u)\big) = 0$. In fact, this is an immediate consequence of the inclusion $Reg(\partial \Omega_u) \subset \partial^* \Omega_u$ and a well-known theorem of Federer, which states that if Ω is a set of finite perimeter, then

$$\mathcal{H}^{d-1}\big(\partial \Omega \setminus (\Omega^{(1)} \cup \Omega^{(0)} \cup \partial^* \Omega)\big) = 0,$$

and of the fact that $\partial \Omega \cap \big(\Omega^{(1)} \cup \Omega^{(0)}\big) = \emptyset$ (see Sect. 5.1). In particular, this completes the proof of claim (ii) of Theorem 1.2.

Chapters 7 and 8 are dedicated to the regularity of $Reg(\partial \Omega_u)$ (Theorem 1.2 (i)). We will use the theory presented in this sections both for Theorem 1.2 and Theorem 1.9.

In Sect. 7.1 (Proposition 7.1) we use the examples of radial solutions from Sect. 2.4 (Propositions 2.15 and 2.16) as test functions to prove that the minimizer u satisfies the following optimality condition in viscosity sense:

$$|\nabla u| = \sqrt{\Lambda} \quad \text{on} \quad \partial \Omega_u \cap D.$$

The Sects. 7.2, 7.3 and 7.4 are dedicated to the proof of the improvement-of-flatness theorem of De Silva [23] (Theorem 7.4), which holds for viscosity solutions. We notice that in the two-dimensional case (Theorem 1.10) all the result from this section will be replaced by the epiperimetric inequality approach from Chap. 12.

In Chap. 8 we show how the improvement of flatness implies the regularity of the free boundary. Precisely, in Sect. 8.1 we prove that the improvement of flatness (Condition 8.3) implies the uniqueness of the blow-up limit u_{x_0} at every point x_0 of the free boundary. Moreover, it provides us with a rate of convergence of the blow-up sequence (Lemma 8.4). Finally, in Sect. 8.2, we show how the uniqueness of the blow-up limit and the rate of convergence of the blow-up sequence imply the $C^{1,\alpha}$ regularity of the free boundary (Proposition 8.6), which concludes the proof of Theorem 1.2.   □

Remark 1.3 The proof of the regularity of $Reg(\partial \Omega_u)$ is based on an improvement-of-flatness argument and is due to De Silva (see [23]). Just as the original proof of Alt and Caffarelli it is based on comparison arguments and does not make use of any type of monotonicity formula. In order to keep the original spirit of [3], we do not use monotonicity formulas in the proof of Theorem 1.2 (Chaps. 2–8). On the other hand, without a monotonicity formula, one can prove that the singular set has zero $(d - 1)$-dimensional Hausdorff measure. Notice that, in [3] it was also shown that the singular set is empty in dimension two. We postpone this result to Sect. 9.4 since it is a trivial consequence of the monotonicity formula of Weiss. We also notice that the proof of Theorem 1.2 is essentially self-contained and requires only basic knowledge on Sobolev spaces and elliptic PDEs.

1.4   The Dimension of the Singular Set

In Theorem 1.2, we show that the singular part of the free boundary $Sing(\partial\Omega_u)$ has the following properties:

- it is a closed subset of the free boundary $\partial\Omega_u \cap D$;
- it has zero Hausdorff measure, that is, $\mathcal{H}^{d-1}\big(Sing(\partial\Omega_u)\big) = 0$; in particular, this implies that the (Hausdorff) dimension of $Sing(\partial\Omega_u)$ is at most $d - 1$.

In [52], using a monotonicity formula and the Federer's dimension reduction principle, Weiss proved the following result.

Theorem 1.4 (Weiss) *Let D be a bounded open set in $\mathbb{R}^d$ and $u \in H^1(D)$ be a non-negative minimizer of $\mathcal{F}_\Lambda$ in D. Let $Reg(\partial\Omega_u)$ and $Sing(\partial\Omega_u)$ be the regular and singular sets from Theorem 1.2. There exists a critical dimension d^* (see Definition 1.5) such that the following holds.*

- *(i) If $d < d^*$, then $Sing(\partial\Omega_u)$ is empty.*
- *(ii) If $d = d^*$, then the singular set $Sing(\partial\Omega_u)$ is a discrete (locally finite) set of isolated points in D.*
- *(iii) If $d > d^*$, then the singular set $Sing(\partial\Omega_u)$ is a closed set of Hausdorff dimension $d - d^*$, that is,*

$$\mathcal{H}^{d-d^*+\varepsilon}(\partial\Omega_u \cap D) = 0 \quad \text{for every} \quad \varepsilon \in (0,1).$$

Definition 1.5 (Definition of d^*) We will denote by d^* the smallest dimension d such that there exists a function $z : \mathbb{R}^d \to \mathbb{R}$ with the following properties:

- z is non-negative and one-homogeneous;
- z is a local minimizer of $\mathcal{F}_\Lambda$ in $\mathbb{R}^d$;
- the free boundary $\partial\Omega_z$ is not a $(d - 1)$-dimensional C^1-regular manifold in $\mathbb{R}^d$.

Remark 1.6 The value of d^* does not depend on $\Lambda > 0$. Without loss of generality, we may take $\Lambda = 1$.

Remark 1.7 (On the Critical Dimension d^*) In this notes, we prove that $d^* \geq 3$ (see Sect. 9.4). Already this is a better estimate (on the dimension of the singular set) with respect to the one from Theorem 1.2 as it means that

$$\mathcal{H}^{d-3+\varepsilon}(\partial\Omega_u \cap D) = 0 \quad \text{for every} \quad \varepsilon \in (0,1).$$

In fact, it is now known that

$$d^* = 5, 6, \text{ or } 7.$$

Fig. 1.3 The free boundary
(in red) of the
one-homogeneous global
solution $u : \mathbb{R}^7 \to \mathbb{R}$ of De
Silva and Jerison

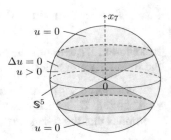

Precisely, Caffarelli, Jerison and Kenig [16] proved that there are no singular one-homogeneous global minimizers in $\mathbb{R}^3$ (thus, $d^* \geq 4$). Later, Jerison and Savin [37] proved the same result in $\mathbb{R}^4$ (so, $d^* \geq 5$). On the other hand, De Silva and Jerison [24] gave an explicit example (see Fig. 1.3) of a singular free boundary in dimension seven (which means that $d^* \leq 7$).

In order to prove Theorem 1.4 we will need most of the theory developed for the proof of Theorem 1.2. For instance, the Lipschitz continuity and the non-degeneracy of the minimizers (Chaps. 3 and 4), the convergence of the blow-up sequences (Chap. 6) and the epsilon regularity theorem (Theorem 8.1 from Chap. 8). On the other hand, we will not need the results from Chap. 5.

The main results that we will need for the proof of Theorem 1.4 are contained in Chaps. 9 and 10. Chapter 9 is dedicated to the Weiss monotonicity formula from [52], which we prove both for minimizing and stationary free boundaries. Chapter 10 is dedicated to the Federer's dimension reduction principle (see [32]). Even if the results of this section concern the one-phase free boundaries, the underlying principle is universal and can be applied to numerous other problems; for instance, in geometric analysis (see [32] and [48]) or to other free boundary problems [42].

Proof of Theorem 1.4 We will first prove that all the blow-up limits of u (at any point of the free boundary) are one-homogeneous global minimizers of $\mathcal{F}_\Lambda$. The global minimality (see Definition 2.12) of the blow-up limits follows from Proposition 6.2. In order to prove the one-homogeneity of the blow-up limits (Proposition 9.12) we will use the Weiss' boundary adjusted energy, defined for any function $\varphi \in H^1(B_1)$ as

$$W_\Lambda(\varphi) := \int_{B_1} |\nabla\varphi|^2 \, dx - \int_{\partial B_1} \varphi^2 \, d\mathcal{H}^{d-1} + \Lambda \big|\{\varphi > 0\} \cap B_1\big|.$$

Let now $x_0 \in \partial\Omega_u \cap D$ and $u_{x_0,r}$ be the usual rescaling (blow-up sequence)

$$u_{r,x_0}(x) = \frac{1}{r} u(x_0 + rx).$$

If we choose $r > 0$ small enough, then the function $u_{x_0,r}$ is defined on B_1 and so, we can compute the Weiss energy $W_\Lambda(u_{x_0,r})$. In Lemma 9.2 we compute the derivative of $W_\Lambda(u_{x_0,r})$ with respect to r, from which we deduce that (see Proposition 9.4):

- the function $r \mapsto W_\Lambda(u_{x_0,r})$ is monotone increasing in r;
- and is constant on an interval of the form $(0, R)$, if and only if, u is one-homogeneous in the ball $B_R(x_0)$.

In particular, the monotonicity of $r \mapsto W_\Lambda(u_{x_0,r})$ and the Lipschitz continuity of u (which gives a lower bound on $W_\Lambda(u_{x_0,r})$) imply that the limit

$$L := \lim_{r \to 0} W_\Lambda(u_{x_0,r}),$$

exists and is finite.

Let now v be a blow-up limit of u at x_0 and $(r_n)_n$ be an infinitesimal sequence such that

$$v = \lim_{n \to \infty} u_{x_0,r_n}.$$

Let $s > 0$ be fixed. Then, the blow-up sequence $u_{x_0,sr_n} = \frac{1}{sr_n}u(x_0 + sr_n x)$ converges locally uniformly to the rescaling $v_s(x) := \frac{1}{s}v(sx)$ of the blow-up v. Now, Proposition 6.2 implies that:

- the sequence u_{x_0,sr_n} converges to v_s strongly in $H^1(B_1)$;
- the sequence of characteristic functions $\mathbb{1}_{\{u_{x_0,sr_n} > 0\}}$ converges to the characteristic function $\mathbb{1}_{\{v_s > 0\}}$ in $L^1(B_1)$.

Thus, for every $s > 0$, we have

$$L = \lim_{r \to 0} W_\Lambda(u_{x_0,r}) = \lim_{n \to \infty} W_\Lambda(u_{x_0,sr_n}) = W_\Lambda(v_s),$$

and so the function $s \mapsto W_\Lambda(v_s)$ is constant in s. Applying again Proposition 9.4, we get that v is one-homogeneous.

Theorem 1.4 now follows by the more general result proved in Proposition 10.13, which can be applied to u since we have the epsilon regularity theorem (Theorem 8.1), the non-degeneracy of u (see Chap. 4), the strong convergence of the blow-up sequences (Proposition 6.2) and the homogeneity of the blow-up limits, which we proved above.   □

Remark 1.8 Finally, we notice that an even better result was recently obtained by Edelen and Engelstein (see [27]). Using the powerful method of Naber and Valtorta (see [44]), they proved that the singular set $Sing(\partial\Omega_u)$ has locally finite $(d - d^*)$—Hausdorff measure, which in particular implies claim (ii) of Theorem 1.4.

1.5 Regularity of the Free Boundary for Measure Constrained Minimizers

Let $D \subset \mathbb{R}^d$ be a smooth and connected bounded open set, $m \in (0, |D|)$ and $g : D \to \mathbb{R}$ be a given non-negative function in $H^1(D)$. This section is dedicated to the following variational minimization problem with measure constraint

$$\min \left\{ \mathcal{F}_0(v, D) \ : \ v \in H^1(D), \ v - g \in H_0^1(D), \ |\Omega_v| = m \right\}, \qquad (1.6)$$

which means

Find $u \in H^1(D)$ such that $u - g \in H_0^1(D)$, $|\Omega_u| = m$ and

$$\mathcal{F}_0(u, D) \leq \mathcal{F}_0(v, D), \ \text{for every } v \in H^1(D) \ \text{such that } v - g \in H_0^1(D) \ \text{and } |\Omega_v| = m.$$

This is the constrained version of the variational problem from Theorems 1.2 and 1.4. We notice that if u is a minimizer of $\mathcal{F}_\Lambda$ in D, for some $\Lambda > 0$, then u is (obviously) a solution to the minimization problem (1.6) with $m := |\Omega_u|$. Conversely, if u is a solution to the variational problem (1.6), then (as we will show in Proposition 11.2) there is a Lagrange multiplier $\Lambda > 0$, depending on u, such that u formally satisfies the optimality condition

$$\Delta u = 0 \quad \text{in} \quad \Omega_u, \quad |\nabla u| = \sqrt{\Lambda} \quad \text{on} \quad \partial \Omega_u \cap D, \qquad (1.7)$$

in the sense that u is stationary for $\mathcal{F}_\Lambda$ in D (see Definition 9.7). Unfortunately, this does not imply that u is a minimizer of $\mathcal{F}_\Lambda$ in D. The free boundary regularity theory for the solutions to (1.6) is more involved since the competitors used to prove the Lipschitz continuity (Chap. 3), non-degeneracy (Chap. 4), improvement of flatness (Chap. 7) and the monotonicity formula (Proposition 9.4) do not satisfy the measure constraint in (1.6).

The free boundary regularity for solutions of (1.6) was first obtained by Aguilera, Alt and Caffarelli in [1]. Our approach is different and strongly relies on the Weiss' monotonicity formula, from which we will deduce both:

- the optimality condition in (1.7) in viscosity sense, which in turn allows to apply the De Silva epsilon regularity theorem (Theorem 8.1) and thus to obtain the $C^{1,\alpha}$-regularity of $Reg(\partial \Omega_u)$ (see Chap. 8);
- the estimate of the dimension of the singular set, which is a consequence of the homogeneity of the blow-up limits and the Federer's dimension reduction (Chap. 10).

Our approach is inspired by the theory developed in [46] and contains several ideas from [41] and the work of Briançon [5] and Briançon-Lamboley [6]. Our main result is the following.

Theorem 1.9 (Regularity of the Measure Constrained Minimizers) *Let D be a connected smooth bounded open set in $\mathbb{R}^d$, $m \in (0, |D|)$ be a positive real constant and $g : D \to \mathbb{R}$ be a given non-negative function in $H^1(D)$. Then, there is a solution to the problem (1.6). Moreover, every solution u is non-negative and locally Lipschitz continuous in D, the set $\Omega_u = \{u > 0\}$ is open and the free boundary can be decomposed as:*

$$\partial\Omega_u \cap D = Reg(\partial\Omega_u) \cup Sing(\partial\Omega_u),$$

where $Reg(\partial\Omega_u)$ and $Sing(\partial\Omega_u)$ are disjoint sets such that:

(i) the regular part $Reg(\partial\Omega_u)$ is a $C^{1,\alpha}$-regular $(d-1)$-dimensional manifold in D, for some $\alpha > 0$;
(ii) the singular part $Sing(\partial\Omega_u)$ is a closed set of Hausdorff dimension $d - d^*$ (where the critical dimension d^* is again given by Definition 1.5), that is,

$$\mathcal{H}^{d-d^*+\varepsilon}(\partial\Omega_u \cap D) = 0 \quad \text{for every} \quad \varepsilon \in (0, 1).$$

Moreover, if $d < d^*$, then $Sing(\partial\Omega_u)$ is empty, and if $d = d^*$, then $Sing(\partial\Omega_u)$ is a countable discrete (locally finite) set of points in D.

Proof of Theorem 1.4 We prove the existence of a solution $u : D \to \mathbb{R}$ in Sect. 11.1, where we also show that u is harmonic in Ω_u in the following sense

$$\int_D |\nabla u|^2 \, dx \leq \int_D |\nabla v|^2 \, dx \quad \text{for every} \quad v \in H^1(D)$$

$$\text{such that} \quad u - v \in H_0^1(D) \quad \text{and} \quad v = 0 \quad \text{on} \quad D \setminus \Omega_u.$$

In particular, applying Lemma 2.7, we get that u is subharmonic in D. Thus, we can suppose that u is defined at every point of D and that

$$u(x_0) := \fint_{\partial B_r(x_0)} u \, d\mathcal{H}^{d-1} = \fint_{B_r(x_0)} u \, dx \quad \text{for every} \quad x_0 \in D.$$

Moreover, the subharmonicity of u implies that it is locally bounded so, from now on, without loss of generality, we will assume that $u \in L^\infty(D)$. Finally, we notice that the set Ω_u is defined everywhere in D (not just up to a set of zero Lebesgue measure) and its topological boundary coincides with the measure-theoretic one (see Lemma 2.9). Precisely, this means that

$$x_0 \in \partial\Omega_u \quad \text{if and only if} \quad 0 < |\Omega_u \cap B_r(x_0)| \leq |B_r| \quad \text{for every} \quad r > 0.$$

In order to prove the Lipschitz continuity of u and the regularity of the free boundary $\partial\Omega_u \cap D$ we proceed in several steps. Notice that we cannot apply directly the results from Chaps. 3–10 since it is not a priori known if the solution u is a local minimizer of $\mathcal{F}_\Lambda$ for some $\Lambda > 0$, that is, one cannot remove the constraint in (1.6) by adding a Lagrange multiplier Λ directly in the functional. In fact, it is only possible to prove the existence of Λ for which the solution u of (1.6) is *stationary* but not minimal for $\mathcal{F}_\Lambda$. From this, we will deduce that u satisfies a quasi-minimality condition, which will allow to proceed as in the proof of Theorems 1.2 and 1.4.

Step 1 *Existence of a Lagrange multiplier.* In Sect. 11.2, we show that there exists $\Lambda > 0$ such that u is stationary for the functional $\mathcal{F}_\Lambda$, that is,

$$\partial \mathcal{F}_\Lambda(u, D)[\xi] = 0 \quad \text{for every} \quad \xi \in C_c^\infty(D; \mathbb{R}^d),$$

where the first variation $\partial \mathcal{F}_\Lambda(u, D)[\xi]$ of $\mathcal{F}_\Lambda$ in the direction of the (compactly supported) vector field ξ is defined in (9.6). The existence of a non-negative Lagrange multiplier can be obtained by a standard variational argument (see Proposition 11.2 and its proof in Sect. 11.2, after Lemma 11.3). The strict positivity of Λ is a non-trivial question which requires some fine analysis of the functions, which are stationary for $\mathcal{F}_0$; we prove it in Sect. 11.3 using the Almgren's frequency function and following the proof of an analogous result from [46], which is a (small with respect to the original result) improvement of the unique continuation principle of Garofalo-Lin [34].

Step 2 *Almost-minimality of u.* Let $x_0 \in \partial \Omega_u \cap D$. In Sect. 11.5 (Proposition 11.10), we prove that u is an almost minimizer of $\mathcal{F}_\Lambda$ (Λ is the Lagrange multiplier from the previous step) in a neighborhood of x_0 in the following sense. There exists a ball B, centered in x_0, in which u satisfies the following almost-minimality condition:

For every $\varepsilon > 0$, there is $r > 0$ such that, for every ball $B_r(y_0) \subset B$, u satisfies the following optimality conditions in $B_r(y_0)$:

$$\mathcal{F}_{\Lambda+\varepsilon}(u, D) \le \mathcal{F}_{\Lambda+\varepsilon}(v, D) \text{ for every } v \in H^1(D) \text{ such that } \begin{cases} v - u \in H_0^1(B_r(y_0)), \\ |\Omega_u| \le |\Omega_v|. \end{cases}$$

$$(1.8)$$

$$\mathcal{F}_{\Lambda-\varepsilon}(u, D) \le \mathcal{F}_{\Lambda-\varepsilon}(v, D) \text{ for every } v \in H^1(D) \text{ such that } \begin{cases} v - u \in H_0^1(B_r(y_0)), \\ |\Omega_u| \ge |\Omega_v|. \end{cases}$$

$$(1.9)$$

The proof of Proposition 11.10 follows step-by-step the proof of the analogous result from [46] and is based on the method of Briançon [5]. Once we have Proposition 11.10, we can proceed as in Theorems 1.2 and 1.4.

Step 3 *Lipschitz continuity and non-degeneracy of u.* In order to prove the (local) Lipschitz continuity of u, we notice that (1.8) leads to an outwards optimality condition. In fact, fixed $\varepsilon > 0$ and $x_0 \in D$, there is a ball $B_r(x_0)$ such that:

$$\mathcal{F}_{\Lambda+\varepsilon}(u, D) \le \mathcal{F}_{\Lambda+\varepsilon}(v, D) \text{ for every } v \in H^1(D) \text{ such that } \begin{cases} v - u \in H_0^1(B_r(x_0)), \\ \Omega_u \subset \Omega_v. \end{cases}$$

$$(1.10)$$

Now, the Lipschitz continuity of u follows by (1.10) and Theorem 3.2.

On the other hand, for the non-degeneracy of u, we notice that, (1.9) implies the following inwards optimality condition:

Fixed $\varepsilon > 0$ and $x_0 \in D$, there is a ball $B_r(x_0)$ such that:

$$\mathcal{F}_{\Lambda-\varepsilon}(u, D) \le \mathcal{F}_{\Lambda-\varepsilon}(v, D) \text{ for every } v \in H^1(D) \text{ such that } \begin{cases} v - u \in H_0^1(B_r(x_0)), \\ \Omega_u \supset \Omega_v. \end{cases}$$

$$(1.11)$$

The non-degeneracy of u follows by (1.11) and the results from Chap. 4 (one can apply both Lemma 4.4 and 4.5).

As a consequence of the Lipschitz continuity and the non-degeneracy of u, we obtain the following results:

- Ω_u satisfies interior and exterior density estimates (Lemma 5.1);
- Ω_u has locally finite perimeter in D (Proposition 5.3);
- $\partial \Omega_u$ has locally finite $(d-1)$-dimensional Hausdorff measure in D (Proposition 5.7).

Step 4. *Convergence of the blow-up sequences and analysis of the blow-up limits.* We recall that, for any $x_0 \in D$ and any $r > 0$, the function

$$u_{x_0,r}(x) := \frac{1}{r} u(x_0 + rx),$$

is well-defined on the set $\frac{1}{r}(-x_0 + D)$ and, in particular, on the ball of radius $\frac{1}{r} \text{dist}(x_0, \partial D)$ centered in zero. By the Lipschitz continuity of u, we notice that for any $x_0 \in \partial \Omega_u \cap D$ and any $R > 0$ the family of functions

$$\left\{ u_{x_0,r} \; : \; 0 < r < \frac{1}{R} \text{dist}(x_0, \partial D) \right\},$$

is equicontinuous and uniformly bounded on the ball $\overline{B}_R \subset \mathbb{R}^d$. This implies that for every sequence u_{x_0,r_n}, with $x_0 \in \partial \Omega_u \cap D$ and $\lim_{n \to \infty} r_n = 0$, there are a subsequence (still denoted by $(u_{x_0,r_n})_{n \in \mathbb{N}}$) and a (Lipschitz) function $u_0 : \mathbb{R}^d \to \mathbb{R}$ such that, for every fixed $R > 0$, the sequence u_{x_0,r_n} converges uniformly to u_0 in the ball

B_R. We say that u_0 is a blow-up limit of u at x_0 and u_{x_0,r_n} is a blow-up sequence. Recall that u is Lipschitz continuous, non-degenerate, harmonic in Ω_u and satisfies the following quasi-minimality condition, which is a direct consequence of (1.8) and (1.9). For every $x_0 \in \partial\Omega_u \cap D$, there are $r_0 > 0$ and a continuous non-negative function $\varepsilon : [0, r] \to \mathbb{R}$, vanishing in zero and such that

$$\mathcal{F}_\Lambda(u, D) \le \mathcal{F}_\Lambda(v, D) + \varepsilon(r)|B_r| \quad \text{for every} \quad 0 < r \le r_0$$

$$\text{and every} \quad v \in H^1(D) \quad \text{such that} \quad u - v \in H_0^1(B_r(x_0)).$$

Let u_{x_0,r_n} be a blow-up sequence converging locally uniformly to the blow-up limit u_0. By Proposition 6.2 and the results of Sect. 6.2 we have that, for every $R > 0$,

(i) the sequence u_{x_0,r_n} converges to u_0 strongly in $H^1(B_R)$;
(ii) the sequence of characteristic functions $\mathbb{1}_{\Omega_n}$ converges to $\mathbb{1}_{\Omega_0}$ in $L^1(B_R)$, where

$$\Omega_n := \{u_{x_0,r_n} > 0\} \quad \text{and} \quad \Omega_0 := \{u_0 > 0\};$$

(iii) the sequence of sets $\overline{\Omega}_n$ converges locally Hausdorff in B_R to $\overline{\Omega}_0$;

Moreover, using again Proposition 6.2, we get that every blow-up limit u_0 of u is a global minimizer of $\mathcal{F}_\Lambda$. Next, since u is a critical point of $\mathcal{F}_\Lambda$, we can apply Lemma 9.11 obtaining that every blow-up limit of u_0 is one-homogeneous. We summarize this in the following statement, with which we conclude this step of the proof:

$$\text{Every blow-up of } u \text{ is a one-homogeneous global minimizer of } \mathcal{F}_\Lambda. \quad (1.12)$$

Step 5. *Optimality condition on the free boundary.* Using the convergence of the blow-up sequences (proved in the previous step) and the structure of the blow-up limits (claim (1.12)), we can apply Proposition 9.18. Thus, u is a viscosity solution of

$$\Delta u = 0 \quad \text{in} \quad \Omega_u, \qquad |\nabla u| = \sqrt{\Lambda} \quad \text{on} \quad \partial\Omega_u \cap D. \qquad (1.13)$$

in viscosity sense (see Definition 7.6).

Step 6. *Decomposition of the free boundary into a regular and a singular parts.* As in the proof of Theorem 1.2, we say that $x_0 \in Reg(\partial\Omega_u)$ if $x_0 \in \partial\Omega_u \cap D$ and there exists a blow-up limit u_0 of u (at x_0), for which there is a unit vector $\nu \in \mathbb{R}^d$ such that

$$u_0(x) = \sqrt{\Lambda}\,(x \cdot \nu)_+ \quad \text{for every} \quad x \in \mathbb{R}^d.$$

The singular part of the free boundary is defined as $Sing(\partial\Omega_u) := (\partial\Omega_u \cap D) \setminus Reg(\partial\Omega_u)$. The $C^{1,\alpha}$-regularity of $Reg(\partial\Omega_u)$ now follows by Theorem 8.1 and the

fact that u is a solution of (1.13). The estimate on the dimension of the singular set
(Theorem 1.9 (ii)) now follows directly from Proposition 10.13.   □

1.6 An Epiperimetric Inequality Approach to the Regularity of the Free Boundary in Dimension Two

This section is dedicated to a recent alternative approach to the regularity of the
free boundaries based on the so-called *epiperimetric inequality*, which was first
introduced by Reifenberg in the contact of area-minimizing surfaces, but in the
context of the one-phase problem, it was first proved in [49]. We restrict our
attention to the two-dimensional case since the epiperimetric inequality is (for
now) known to hold only in dimension two (see Theorem 12.1 and Theorem 12.3).
Precisely, we will give an alternative proof to the following result.

Theorem 1.10 (Regularity of the Free Boundary in Dimension Two) *Let D be a
bounded open set in $\mathbb{R}^2$. Let $u : D \to \mathbb{R}$ be a non-negative function and a minimizer
of $\mathcal{F}_\Lambda$ in D. Then:*

(i) u is locally Lipschitz continuous in D and the set $\Omega_u = \{u > 0\}$ is open;
(ii) the free boundary $\partial \Omega_u \cap D$ is $C^{1,\alpha}$-regular.

Proof of Theorem 1.4 We first notice that the Lipschitz continuity of u follows by
Theorem 3.1. In what follows, without loss of generality we assume that $\Lambda = 1$. By
the non-degeneracy of the solutions (Chap. 4) and the convergence of the blow-up
sequences (Chap. 6, Proposition 6.2), we have that, for every free boundary point
$x_0 \in \partial \Omega_u$ and every infinitesimal sequence $r_n \to 0$, there exists a subsequence of
r_n (still denoted by r_n) such that u_{x_0,r_n} converges locally uniformly to a non-trivial
blow-up limit $u_0 : \mathbb{R}^2 \to \mathbb{R}$. Moreover,

- the sequence u_{x_0,r_n} converges to u_0 strongly in $H^1(B_1)$;
- the sequence of characteristic functions $\mathbb{1}_{\{u_{x_0,r_n}>0\}}$ converge to $\mathbb{1}_{\{u_0>0\}}$ in $L^1(B_1)$.

Next, we notice that by the Weiss monotonicity formula (Proposition 9.4) the
function $r \mapsto W_1(u_{x_0,r})$ is monotone increasing in r and the blow-up limit u_0
is one-homogeneous global minimizer of $\mathcal{F}_1$ in $\mathbb{R}^2$ (see Lemma 9.10). Thus, by
Proposition 9.13, we obtain that u_0 is a half-plane solution, that is

$$u_0(x) = (x \cdot \nu)_+ ,$$

for some unit vector $\nu \in \mathbb{R}^2$. Now, the strong convergence of the blow-up sequence
and the monotonicity formula (Proposition 9.4) imply that

$$\inf_{r>0} W_1(u_{x_0,r}) = \lim_{r \to 0} W_1(u_{x_0,r}) = \lim_{n \to \infty} W_1(u_{x_0,r_n}) = W_1(u_0) = \frac{\pi}{2}.$$

In conclusion, we have that:

- the energy

$$\mathcal{E}(u) := W_1(u) - \frac{\pi}{2},$$

  is non-negative along any blow-up sequence $u_{x_0,r}$ with $x_0 \in \partial\Omega_u \cap D$,

$$\mathcal{E}(u_{x_0,r}) := W_1(u_{x_0,r}) - \frac{\pi}{2} \geq 0 \quad \text{for every} \quad r > 0;$$

- the free boundary is *flat*, that is, for every $x_0 \in \partial\Omega_u \cap D$ and every $\varepsilon > 0$, there exists $r > 0$ and $\nu \in \partial B_1$, such that:

$$(x \cdot \nu - \varepsilon)_+ \leq u_{x_0,r}(x) \leq (x \cdot \nu + \varepsilon)_+ \quad \text{for every} \quad x \in B_1.$$

Now, by the epiperimetric inequality (Theorem 12.1) and Proposition 12.13, we obtain that, in a neighborhood of x_0, $\partial\Omega_u$ is the graph of a $C^{1,\alpha}$ regular function.

$\square$

1.7   Further Results

The main objective of these notes is to introduce the reader to the free boundary regularity theory and to provide a complete and self-contained proof of the regularity of the one-phase free boundaries. In this perspective, our main results are Theorems 1.2, 1.4, 1.9 and 1.10. On the other hand, in these notes, we also prove several other results, which might be interesting for specialists and non. Here is a list of results, by section, which are worth to be mentioned in this context.

Chapter 2   In Proposition 2.10 we give a direct proof to the fact that the half-plane solutions are global minimizers of $\mathcal{F}_\Lambda$. This is well-known, as the result can be obtained from the following facts:

- the blow-up limits of a solution u at points of the reduced boundary $\partial^*\Omega_u$ are half-plane solutions (Lemma 6.11);
- the reduced boundary $\partial^*\Omega_u$ is non-empty as Ω_u is a set of finite perimeter (Proposition 5.3) and for sets of finite perimeter we have the identity $Per(\Omega_u) = \mathcal{H}^{d-1}(\partial^*\Omega_u)$ (see [43]).

In Lemmas 2.15 and 2.16 we prove the existence and the uniqueness of two one-phase free boundary problems. Moreover, we prove that the solutions are radially symmetric and we write them explicitly.

Chapters 3 and 4 In Proposition 3.15 and Lemma 4.5, we present the methods of Danielli-Petrosyan ([18], for the Lipschitz continuity) and David-Toro ([19], for the non-degeneracy) in the simplified context of the classical one-phase Bernoulli problem. Both methods are very robust and can be applied to more general free boundary problems.

Chapter 5 In Proposition 5.3 we prove that if u is a minimizer of $\mathcal{F}_\Lambda$ in a set D, then Ω_u has locally finite De Giorgi perimeter in D. The method is a localized version of a global estimate by Bucur (see [8]), on the perimeter of the optimal shapes for the eigenvalues of the Dirichlet Laplacian.

In Proposition 5.7 we prove that, if u is a minimizer of $\mathcal{F}_\Lambda$ in a set D, then the $\mathcal{H}^{d-1}$ Hausdorff measure of the free boundary $\partial\Omega_u$ is locally finite in D. The method is very general and can be applied to many different free boundary problems, for instance, to the vectorial problem (see [42]).

Chapter 6 In Proposition 6.2 we give the detailed proof of the strong convergence of the blow-up sequences, which is often omitted in the literature. Moreover, we state and prove a general result (Lemma 6.3) which can be applied to different free boundary and shape optimization problems.

Chapter 7 In Proposition 7.1 we prove that if u is a minimizer of $\mathcal{F}_\Lambda$ in D, then it is satisfies the optimality condition

$$|\nabla u| = \sqrt{\Lambda} \quad \text{on} \quad \partial\Omega_u \cap D,$$

in viscosity sense (Definition 7.6). This result is well-known, but in the literature the proof is usually omitted. Our proof of Proposition 7.1 is based on a comparison with the radial solutions constructed in Lemmas 2.15 and 2.16. We give another proof of this fact in Chap. 9.

Chapter 8 In this section we give a detailed proof of the fact that the improvement of flatness (Condition 8.3) implies the $C^{1,\alpha}$ regularity of the free boundary (see Lemma 8.4 and Proposition 8.6). In particular, in Sect. 8.2, we explain the relation between the uniqueness of the blow-up limits, the rate of convergence of the blow-up sequences, and the regularity of the free boundary (Proposition 8.6).

Chapter 9 In Sect. 9.5, we give another proof of the fact that, if u is a local minimizer of $\mathcal{F}_\Lambda$ in D, then it satisfies the optimality condition

$$|\nabla u| = \sqrt{\Lambda} \quad \text{on} \quad \partial\Omega_u \cap D,$$

in viscosity sense (see also Proposition 7.1). The method that we propose is based on the Weiss monotonicity formula and is very robust, for instance, it applies to general operators (see [46]) and to vectorial problems (see [41]). This method was first introduced in [41].

Chapter 10 This section is an introduction to the Federer's Dimension Reduction Principle in the context of free boundary problems. Our main result (Proposition 10.13) is an estimate on the dimension of the singular set under general conditions.

Chapter 11 In Sect. 11.3 we combine the unique continuation principle of Garofalo-Lin [34] with the Faber-Krahn-type inequality from [10] to prove a strong unique continuation result for stationary functions of the Dirichlet energy $\mathcal{F}_0$ (see Proposition 9.19 and [46]).

Chapter 12 This section is dedicated to the epiperimetric inequality (Theorem 12.1) that first appeared in [49]. We give here a different proof that inspired the approach to the epiperimetric inequality at the singular points in higher dimension (see [29]).

In Lemma 12.14 we prove that the epiperimetric inequality at the flat free boundary points in any dimension (Condition 12.12) implies the regularity of the free boundary. The proof is similar to the one in [49], but has to deal with the closeness condition in the epiperimetric inequality (see Condition 12.12), precisely as in [29] and [28].

In Sect. 12.6 we prove comparison results for minimizers of $\mathcal{F}_\Lambda$ (Proposition 12.19 and Lemma 12.22) and for viscosity solutions (Lemma 12.21).

In Theorem 12.3 we prove an epiperimetric inequality in dimension two without any specific assumption on the trace on the sphere. This results covers both Theorem 12.1 and the main theorem of [49]. Both Theorem 12.3 and Theorem 12.1 are new results.

Chapter 2
Existence of Solutions, Qualitative Properties and Examples

In this section, we prove that local minimizers of the functional $\mathcal{F}_\Lambda$ do exist (Proposition 2.1) and we give several important examples of local minimizers that can be computed explicitly (Proposition 2.10, Lemmas 2.15 and 2.16).

Proposition 2.1 *Let $\Lambda > 0$, $D \subset \mathbb{R}^d$ be a bounded open set and the function $g \in H^1(D)$ be fixed and such that $g \geq 0$ in D. Then, there exists a solution to the variational problem*

$$\min\big\{\mathcal{F}_\Lambda(u, D) \; : \; u \in H^1(D), \, u - g \in H^1_0(D)\big\}. \tag{2.1}$$

Moreover, every solution u of (2.1) has the following properties:

 (i) *u is non-negative in D;*
 (ii) *u is locally bounded in D;*
 (iii) *there is a function $\tilde{u} : D \to \mathbb{R}$ such that $\tilde{u} \geq 0$ and $\tilde{u} = u$ almost everywhere in D and*

$$\tilde{u}(x_0) = \lim_{r \to 0} \frac{1}{|B_r|} \int_{B_r(x_0)} \tilde{u}(x)\, dx \qquad \text{for every} \qquad x_0 \in D.$$

Remark 2.2 From now on, we will identify any solution u of (2.1) with its representative $\tilde{u}$; for the sake of simplicity, we will always write u instead of $\tilde{u}$.

The rest of the section is organized as follows. In Sect. 2.1 we discuss some of the properties (scaling and truncation) of the function $\mathcal{F}_\Lambda$. Section 2.2 is dedicated to the proof of Proposition 2.1. In Sects. 2.3 and 2.4, we discuss several examples of local minimizers, which we will find application in the next sections.

© The Author(s) 2023
B. Velichkov, *Regularity of the One-phase Free Boundaries*,
Lecture Notes of the Unione Matematica Italiana 28,
https://doi.org/10.1007/978-3-031-13238-4_2

2.1 Properties of the Functional $\mathcal{F}$

In this section, we discuss several basic properties of the functional

$$(\Lambda, t, D) \mapsto \mathcal{F}_\Lambda(u, D).$$

We give the precise statements in Lemmas 2.3, 2.4 and 2.5.

Lemma 2.3 (Scaling) *Let* $\Omega \subset \mathbb{R}^d$ *be an open set and* $u \in H^1(\Omega)$.

(a) Let $x_0 \in \mathbb{R}$, $r > 0$ *and*

$$u_{x_0,r}(x) := \frac{1}{r} u(x_0 + rx) \qquad and \qquad \Omega_{x_0,r} = \left\{ x = \frac{y - x_0}{r} \in \mathbb{R} : y \in \Omega \right\}.$$

Then $u_{x_0,r} \in H^1(\Omega_{x_0,r})$ *and*

$$\mathcal{F}_\Lambda(u_{x_0,r}, \Omega_{x_0,r}) = r^{-d} \, \mathcal{F}_\Lambda(u, \Omega).$$

In particular, if u *is a minimizer of* $\mathcal{F}_\Lambda$ *in* Ω, *then* $u_{x_0,r}$ *is a minimizer of* $\mathcal{F}_\Lambda$ *in* $\Omega_{x_0,r}$.

(b) For every $t > 0$, *we have*

$$\mathcal{F}_{t^2\Lambda}(tu, \Omega) = t^2 \, \mathcal{F}_\Lambda(u, \Omega).$$

In particular, if u *is a minimizer of* $\mathcal{F}_\Lambda$ *in* Ω, *then* tu *is a minimizer of* $\mathcal{F}_{t^2\Lambda}$ *in* Ω.

Proof The proof is a straightforward computation. $\square$

Lemma 2.4 (Truncation) *Let* $\Omega \subset \mathbb{R}^d$ *be an open set and* $u \in H^1(\Omega)$. *Then,*

$$\mathcal{F}_\Lambda(u, \Omega) - \mathcal{F}_\Lambda(0 \vee u, \Omega) = \int_{\{u<0\}\cap\Omega} |\nabla u|^2 \, dx.$$

Moreover, for every $t \geq 0$, *we have*

$$\mathcal{F}_\Lambda(u, \Omega) - \mathcal{F}_\Lambda(u \wedge t, \Omega) = \int_{\{u>t\}\cap\Omega} |\nabla u|^2 \, dx.$$

Proof The proof follows by the definition of $\mathcal{F}$ and the identities

$$\nabla(u \wedge t) = \mathbb{1}_{\{u<t\}} \nabla u \qquad and \qquad \nabla(u \vee 0) = \mathbb{1}_{\{u>0\}} \nabla u.$$

$\square$

Lemma 2.5 (Comparison) *Let $\Omega \subset \mathbb{R}^d$ be an open set and $u, v \in H^1(\Omega)$ be two given functions. Then we have*

$$\mathcal{F}_\Lambda(u \vee v, \Omega) + \mathcal{F}_\Lambda(u \wedge v, \Omega) = \mathcal{F}_\Lambda(u, \Omega) + \mathcal{F}_\Lambda(v, \Omega).$$

Proof The proof is a straightforward computation. In fact, we have

$$\mathcal{F}_\Lambda(u \vee v, \Omega) + \mathcal{F}_\Lambda(u \wedge v, \Omega)$$

$$= \int_\Omega |\nabla(u \vee v)|^2 \, dx + \Lambda |\{u \vee v > 0\} \cap \Omega|$$

$$\quad + \int_\Omega |\nabla(u \wedge v)|^2 \, dx + \Lambda |\{u \wedge v > 0\} \cap \Omega|$$

$$= \int_{\Omega \cap \{u \geq v\}} |\nabla u|^2 \, dx + \int_{\Omega \cap \{u < v\}} |\nabla v|^2 \, dx + \Lambda \big|(\{u > 0\} \cup \{v > 0\}) \cap \Omega\big|$$

$$\quad + \int_{\Omega \cap \{u \geq v\}} |\nabla v|^2 \, dx + \int_{\Omega \cap \{u < v\}} |\nabla u|^2 \, dx + \Lambda |\{u > 0\} \cap \{v > 0\} \cap \Omega|$$

$$= \int_\Omega |\nabla u|^2 \, dx + \Lambda |\{u > 0\} \cap \Omega| + \int_\Omega |\nabla v|^2 \, dx + \Lambda |\{v > 0\} \cap \Omega|$$

$$= \mathcal{F}_\Lambda(u, \Omega) + \mathcal{F}_\Lambda(v, \Omega),$$

which concludes the proof. $\qquad\qquad\qquad\qquad\qquad\qquad\qquad\qquad\qquad\qquad\qquad$ □

2.2 Proof of Proposition 2.1

In this section we prove Proposition 2.1. We will first show that the minimizers of $\mathcal{F}_\Lambda$ are subharmonic functions (Lemmas 2.6 and 2.7) and then we will deduce the claim (iii) of Proposition 2.1 (see Remark 2.2). At the end of this section, we will complete the proof of Proposition 2.1 by proving that there is a solution to the variational problem (2.1). Finally, in Lemma 2.9, we discuss the definition of the free boundary, which can be (equivalently) defined both as the topological boundary of the representative $\tilde{u}$ (of the function $u \in H^1(D)$) defined in Proposition 2.1 and as the measure-theoretic boundary of Ω_u, which does not depend on the representative of u and is defined as the set of points $x_0 \in D$ for which

$$|B_r(x_0) \cap \Omega_u| > 0 \quad \text{and} \quad |\Omega_u \setminus B_r(x_0)| > 0 \quad \text{for every} \quad r > 0.$$

Lemma 2.6 (The Minimizers of $\mathcal{F}_\Lambda$ Are Subharmonic Functions) *Let $D \subset \mathbb{R}^d$ be a bounded open set and the non-negative function $u \in H^1(D)$ be a minimizer of*

$\mathcal{F}_\Lambda$ in D. Then u is subharmonic, $\Delta u \geq 0$, on D in sense of distributions:

$$\int_D \nabla u \cdot \nabla \varphi \, dx \leq 0 \quad \text{for every} \quad \varphi \in C_c^\infty(D) \quad \text{such that} \quad \varphi \geq 0 \quad \text{on} \quad D.$$

Proof Let $\varphi \in C_c^\infty(D)$ be a given non-negative function. Suppose that $t \geq 0$ and $v = u - t\varphi$. Then we have that $v_+ \leq u$. In particular, integrating on the support of φ we have

$$\mathcal{F}_\Lambda(u, D) = \int_D |\nabla u|^2 \, dx + \Lambda |\{u > 0\} \cap D|$$

$$\leq \int_D |\nabla v_+|^2 \, dx + \Lambda |\{v_+ > 0\} \cap D| \leq \int_D |\nabla v|^2 \, dx + \Lambda |\{u > 0\} \cap D|.$$

This implies that

$$\int_D |\nabla u|^2 \, dx \leq \int_D |\nabla (u - t\varphi)|^2 \, dx = \int_D |\nabla u|^2 \, dx - 2t \int_D \nabla u \cdot \nabla \varphi \, dx + t^2 \int_D |\nabla \varphi|^2 \, dx,$$

and the claim follows by taking the (right) derivative at $t = 0$. $\qquad\square$

There is also a more general result, which applies not only to minimizers, but also to generic non-negative functions, which are harmonic where they are strictly positive. The proof can also be found in the book of Henrot and Pierre [36].

Lemma 2.7 (The Minimizers of $\mathcal{F}_\Lambda$ Are Subharmonic Functions II) *Let $D \subset \mathbb{R}^d$ be a bounded open set and the non-negative function $u \in H^1(D)$ be harmonic in the set $\Omega_u := \{u > 0\}$, that is*

$$\int_D |\nabla u|^2 \, dx \leq \int_D |\nabla v|^2 \, dx \quad \text{for every} \quad v \in H^1(D)$$

$$\text{such that} \quad u - v \in H_0^1(D) \quad \text{and} \quad v = 0 \quad \text{on} \quad D \setminus \Omega_u.$$

Then u is subharmonic, $\Delta u \geq 0$, on D in sense of distributions.

Proof Let $\phi \in C_c^\infty(D)$ be a given non-negative function and let $p_\varepsilon : \mathbb{R} \to \mathbb{R}$ be given by

$$p_\varepsilon(x) = \begin{cases} 0 & \text{if } x \leq \varepsilon/2, \\ \dfrac{1}{\varepsilon}(2x - \varepsilon) & \text{if } x \in [\varepsilon/2, \varepsilon], \\ 1 & \text{if } x \geq \varepsilon. \end{cases}$$

Since $u_t := u + t \, p_\varepsilon(u)\phi$ is a competitor for u and for $t \in \mathbb{R}$ small enough

$$\{u > 0\} = \{u_t > 0\},$$

we have that for t small enough

$$\int_D |\nabla u|^2 \, dx \le \int_D |\nabla u_t|^2 \, dx,$$

which gives

$$\int_D p_\varepsilon(u)\nabla u \cdot \nabla\phi \, dx \le \int_D p'_\varepsilon(u)|\nabla u|^2\phi \, dx + \int_D p_\varepsilon(u)\nabla u \cdot \nabla\phi \, dx$$

$$= \int_D \nabla u \cdot \nabla(p_\varepsilon(u)\phi) \, dx = 0,$$

where the last inequality is due to the fact that p_ε is increasing. Now since $p_\varepsilon(u)$ converges to $\mathbb{1}_{\{u>0\}}$, as $\varepsilon \to 0$, we get that

$$\int_D \nabla u \cdot \nabla\phi \, dx \le 0,$$

which concludes the proof.   □

Remark 2.8 (Pointwise Definition of a Subharmonic Function) Let D be an open set and $u \in H^1(D)$ be a subharmonic function. Then, for every $x_0 \in D$, we have that

$$\text{the functions} \quad r \mapsto \fint_{\partial B_r(x_0)} u \, d\mathcal{H}^{d-1} \quad \text{and} \quad r \mapsto \fint_{B_r(x_0)} u \, dx \quad \text{are non-decreasing.}$$

$$(2.2)$$

As a consequence of (2.2), we obtain that:

- u is locally bounded, $u \in L^\infty_{loc}(D)$;
- we define $\tilde{u} : D \to \mathbb{R}$ as

$$\tilde{u}(x_0) := \lim_{r \to 0^+} \fint_{B_r(x_0)} u(x) \, dx \quad \text{for every} \quad x_0 \in D.$$

Proof of Proposition 2.1 We first prove that a solution exists. Let $u_n \in H^1(D)$ be a minimizing sequence such that $u_n - g \in H^1_0(D)$ and

$$\mathcal{F}_\Lambda(u_n, D) \le \mathcal{F}_\Lambda(g, D) \quad \text{for every} \quad n \ge 1.$$

By Lemma 2.4 we may assume that, for every $n \ge 1$, $u_n \ge 0$ on D. For simplicity, we assume that $d > 2$ (the case $d = 2$ is analogous) and we set $2^* = \dfrac{2d}{d-2}$. Then, we have

$$\|u_n - g\|^2_{L^{2^*}(D)} \le C_d \int_D |\nabla(u_n - g)|^2 \, dx \le 2C_d \left(\int_D |\nabla u_n|^2 \, dx + \int_D |\nabla g|^2 \, dx \right)$$

$$\le 2C_d\big(\mathcal{F}_\Lambda(u_n, D) + \mathcal{F}_\Lambda(g, D)\big) \le 4C_d \mathcal{F}_\Lambda(g, D).$$

Now, we estimate,

$$\|u_n - g\|^2_{L^2(D)} \le |\{u_n - g \ne 0\}|^{2/d} \|u_n - g\|^2_{L^{2^*}(D)}$$

$$\le \big(|\{u_n > 0\} \cap D| + |\{g > 0\} \cap D|\big)^{2/d} 4C_d \mathcal{F}_\Lambda(g, D) \le 8C_d \Lambda^{-\frac{2}{d}} \mathcal{F}_\Lambda(g, D)^{\frac{2+d}{d}},$$

which implies that the sequence u_n is uniformly bounded in $H^1(D)$. Then, up to a subsequence, we may assume that u_n converges weakly in $H^1(D)$ and strongly in $L^2(D)$ to a function $u \in H^1(D)$. Now, the semi-continuity of the H^1 norm (with respect to the weak H^1 convergence) gives that

$$\int_D |\nabla u|^2 \, dx \le \liminf_{n \to \infty} \int_D |\nabla u_n|^2 \, dx.$$

On the other hand, passing again to a subsequence, we get that u_n converges pointwise almost everywhere to u. This implies that

$$\mathbb{1}_{\{u>0\}} \le \liminf_{n \to \infty} \mathbb{1}_{\{u_n>0\}},$$

and so,

$$|\{u > 0\} \cap D| \le \liminf_{n \to \infty} |\{u_n > 0\} \cap D|,$$

which finally gives that

$$\mathcal{F}_\Lambda(u, D) \le \liminf_{n \to \infty} \mathcal{F}_\Lambda(u_n, D),$$

and so, u is a solution to (2.1). Now, we notice that Lemma 2.4 implies that $u \ge 0$ on D. Lemma 2.6 and Remark 2.8 give the claims (ii) and (iii).   □

We conclude this subsection with the following lemma, where we show that the set Ω_u has a topological boundary that coincides with the measure theoretic one.

Lemma 2.9 (Topological and Measure Theoretic Free Boundaries) *Let $D \subset \mathbb{R}^d$ be a bounded open set and u be a local minimizer of $\mathcal{F}_\Lambda$ in the open set $D \subset \mathbb{R}^d$ or, more generally, let $u : D \to \mathbb{R}$, $u \in H^1(D)$, be a non-negative function satisfying*

(a) u is harmonic in $\Omega_u = \{u > 0\}$ in the sense that

$$\int_D |\nabla u|^2 \, dx \le \int_D |\nabla v|^2 \, dx \quad \text{for every} \quad v \in H^1(D)$$

$$\text{such that} \quad u - v \in H_0^1(D) \quad \text{and} \quad v = 0 \quad \text{on} \quad D \setminus \Omega_u.$$

(b) u is defined everywhere in D and

$$u(x_0) := \lim_{r \to 0^+} \fint_{B_r(x_0)} u(x) \, dx \quad \text{for every} \quad x_0 \in D.$$

Then, the topological boundary of Ω_u coincides with the measure-theoretic one:

$$\partial \Omega_u \cap D = \Big\{ x \in D \ : \ |B_r(x) \cap \Omega_u| > 0 \quad \text{and} \quad |B_r(x) \cap \{u = 0\}| > 0, \ \forall r > 0 \Big\}.$$

Proof We first notice that the following inclusion holds :

$$\partial \Omega_u \cap D \supset \Big\{ x \in D \ : \ |B_r(x) \cap \Omega_u| > 0 \quad \text{and} \quad |B_r(x) \cap \{u = 0\}| > 0, \ \forall r > 0 \Big\}.$$

In order to prove the opposite inclusion we show that

(i) if $|B_r \cap \{u = 0\}| = 0$, then u is harmonic in B_r and $B_r \cap \{u = 0\} = \emptyset$.
(ii) if $|B_r \cap \{u > 0\}| = 0$, then $u = 0$ in B_r, i.e. $B_r \cap \{u > 0\} = \emptyset$.

In order to prove (i) we notice that u is necessarily harmonic in B_r, since otherwise we can contradict the minimality of u by replacing it with the harmonic function with the same boundary values. By the strong maximum principle, u is strictly positive in B_r. The proof of (ii) follows directly from (b). $\square$

2.3   Half-Plane Solutions

The so-called half-plane solutions (see Fig. 2.1)

$$h_\nu(x) = \sqrt{\Lambda} \, (x \cdot \nu)_+$$

play a fundamental role in the free boundary regularity theory. In fact, in the next sections we will show that if a local minimizer u is close to a half-plane solution (at some, possibly very small, scale), then the free boundary is $C^{1,\alpha}$ regular; then,

Fig. 2.1 A half-plane
solution

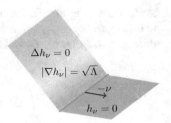

$$\Delta h_\nu = 0$$
$$|\nabla h_\nu| = \sqrt{\Lambda}$$
$$-\nu$$
$$h_\nu = 0$$

we will also prove that at almost-every free boundary point the solution u coincides
with a half-plane solution at order 1.

In this subsection, we make a first step in this direction and we prove that the half-
plane solutions are global minimizers. This result is usually omitted in the literature
since it is implicitly contained in the fact that the blow-up limits at the points of the
reduced free boundary (of any local minimizer) are indeed half-plane solutions (we
will prove this fact later, in Lemma 6.11). The main result of this subsection is the
following.

Proposition 2.10 (The Half-Plane Solutions Are Local Minimizers) *Let $\nu \in \mathbb{R}^d$
be a unit vector. Then the function $H_\nu(x) = \sqrt{\Lambda}\,(\nu \cdot x)_+$ is a global minimizer of
$\mathcal{F}_\Lambda$.*

Definition 2.11 (Local Minimizers) Let D be an open set in $\mathbb{R}^d$. We say that the
function $u : D \to \mathbb{R}$ is a local minimizer of $\mathcal{F}_\Lambda$ in D, if $u \in H^1_{loc}(D)$, $u \geq 0$, and
for any bounded open set Ω such that $\overline{\Omega} \subset D$, we have

$$\mathcal{F}_\Lambda(u, \Omega) \leq \mathcal{F}_\Lambda(v, \Omega) \qquad \text{for every} \qquad v \in H^1_{loc}(D) \quad \text{such that} \quad u - v \in H^1_0(\Omega).$$

Definition 2.12 (Global Minimizers) We say that the function $u : \mathbb{R}^d \to \mathbb{R}$ is a
global minimizer of $\mathcal{F}_\Lambda$, if u is non-negative on $\mathbb{R}^d$, $u \in H^1_{loc}(\mathbb{R}^d)$ and u is a local
minimizer of $\mathcal{F}_\Lambda$ in $\mathbb{R}^d$.

In order to prove the minimality of the half-plane solutions, we will need the
following lemma. We notice that it is useful also in other contexts. For instance, it
allows to prove that the solutions of (2.1) are bounded.

Lemma 2.13 *Let $D \subset \mathbb{R}^d$ be a bounded smooth open set or $D = \mathbb{R}^d$. Let $x_0 \in \mathbb{R}^d$
be a given point, $\nu \in \mathbb{R}^d$ be a unit vector and let*

$$v(x) = h_\nu(x - x_0) = \sqrt{\Lambda}\,\sup\{0, (x - x_0) \cdot \nu\}.$$

Suppose that $u \in H^1(D)$ is a non-negative function such that

$$u = 0 \quad on \quad \partial D \cap \{v = 0\}.$$

Then

$$\mathcal{F}_\Lambda(u \wedge v, D) \leq \mathcal{F}_\Lambda(u, D),$$

with an equality if and only if $u = u \wedge v$.

  In particular, if u is a solution to (2.1), then u has bounded support. Precisely, $u = 0$ outside the set $conv(D) + B_1$, where $conv(D)$ is the convex hull of D.

Proof Without loss of generality we can suppose that $v = e_d$ and $x_0 = 0$. For the sake of simplicity, we set $H_+ = \{x_d > 0\}$ and $H_- = \{x_d < 0\}$. Then

$$\mathcal{F}_\Lambda(u, D) - \mathcal{F}_\Lambda(u \wedge v, D) = \int_{H_-} |\nabla u|^2 \, dx + \Lambda |H_- \cap \{u > 0\}|$$

$$+ \int_{H_+ \cap \{u > \sqrt{\Lambda} x_d\}} \left(|\nabla u|^2 - |\nabla v|^2\right) dx,$$

where (in the case when D is bounded) we assume that u is extended by zero on $H_- \setminus D$. By the fact that $v(x) = \Lambda x_d^+$ is harmonic on $\{x_d > 0\}$, we get that

$$\int_{H_+ \cap \{u > \sqrt{\Lambda} x_d\}} \left(|\nabla u|^2 - |\nabla v|^2\right) dx = \int_{H_+ \cap \{u > \sqrt{\Lambda} x_d\}} \left(|\nabla(u - v)|^2 + 2\nabla v \cdot \nabla(u - v)_+\right) dx$$

$$= \int_{H_+ \cap \{u > \sqrt{\Lambda} x_d\}} |\nabla(u - v)|^2 \, dx - 2\sqrt{\Lambda} \int_{\{x_d = 0\}} u \, d\mathcal{H}^{d-1}.$$

We recall that for every $u \in H^1(\{x_d < 0\})$ we have the inequality[1]

$$\int_{\{x_d < 0\}} |\nabla u|^2 \, dx + \Lambda |\{u > 0\} \cap \{x_d < 0\}| \geq 2\sqrt{\Lambda} \int_{\{x_d = 0\}} u \, d\mathcal{H}^{d-1},$$

where the equality holds, if and only if, $u \equiv 0$ on $\{x_d < 0\}$. Thus, we obtain

$$\mathcal{F}_\Lambda(u, \Omega) - \mathcal{F}_\Lambda(u \wedge v, \Omega) \geq \int_{H_+ \cap \{u > \sqrt{\Lambda} x_d\}} |\nabla(u - v)|^2 \, dx \geq 0,$$

where the last inequality is an equality if and only if $u \leq v$ on $\mathbb{R}^d$.   □

---

[1] Indeed, if $f : \mathbb{R} \to \mathbb{R}^+$ is a Sobolev function such that $f(a) = 0$ for some $a < 0$, then we have

$$f(0) = \int_a^0 f'(t) \, dt \leq |\{f \neq 0\} \cap \{a \leq t \leq 0\}|^{\frac{1}{2}} \left(\int_a^0 |f'(t)|^2 \, dt\right)^{\frac{1}{2}} \leq \frac{1}{2} \left(|\{f \neq 0\} \cap \{t \leq 0\}| + \int_a^0 |f'(t)|^2 \, dt\right).$$

Proof of Proposition 2.10 Without loss of generality we may suppose that $\nu = e_d$ and set

$$h(x) = \sqrt{\Lambda}\, x_d^+.$$

Suppose that $R > 0$ and $u \in H^1_{loc}(\mathbb{R}^d)$ is a non-negative function such that $u - h \in H^1_0(B_R)$. It is sufficient to prove that $\mathcal{F}_\Lambda(h, B_R) \le \mathcal{F}_\Lambda(u, B_R)$. By Lemma 2.13 we have that

$$\mathcal{F}_\Lambda(u \wedge h, B_R) \le \mathcal{F}_\Lambda(u, B_R).$$

Thus, we may suppose that $u \le h$. Since h is harmonic in $\{x_d > 0\}$ we get that

$$\mathcal{F}_\Lambda(u, B_R) - \mathcal{F}_\Lambda(h, B_R) = \int_{\{x_d>0\}} \left|\nabla(u - h)\right|^2 dx - \Lambda\left|\{x_d > 0\} \cap \{u = 0\}\right|$$

$$= \int_{\{x_d>0\}\cap\{u>0\}} \left|\nabla(u - h)\right|^2 dx,$$

where the last equality is due to the fact that

$$|\nabla(u - h)| = |\nabla h| = \sqrt{\Lambda} \qquad \text{on the set} \qquad \{u = 0\}.$$

This concludes the proof. $\square$

2.4   Radial Solutions

In this subsection, we give two examples of local minimizers, which are radial functions. Despite of being ones of the few non-trivial examples of local minimizers, they will also be useful in the proof (to be precise, in one of the two proofs that we will give) of the fact that the local minimizers satisfy an overdetermined condition on the free boundary in viscosity sense (see Definition 7.6 and Proposition 7.1).

Let D be a bounded open set in $\mathbb{R}^d$ with smooth boundary. We consider the following variational minimization problem in the exterior domain $\mathbb{R}^d \setminus \overline{D}$.

$$\min\left\{\int_{\mathbb{R}^d} |\nabla u|^2\, dx + |\{u > 0\}| \; : \; u \in H^1(\mathbb{R}^d),\, u = 1 \text{ in } D\right\}. \tag{2.3}$$

The "interior" version of this problem reads as

$$\min\left\{\int_D |\nabla u|^2\, dx + |\{u > 0\} \cap D| \; : \; u \in H^1(D),\, u = 1 \text{ on } \partial D\right\}. \tag{2.4}$$

We first prove that the problems (2.3) and (2.4) admit solutions.

Lemma 2.14 (Existence of a Solution) *Suppose that D is a bounded open set in $\mathbb{R}^d$ with smooth boundary. Then the variational problems (2.3) and (2.4) admit solutions.*

Proof We give the proof for (2.3), the case (2.4) being analogous (and easier as it does not require the use of Lemma 2.13). Let u_n be a minimizing sequence in $H^1(\mathbb{R}^d)$. By Lemmas 2.4 and 2.13 we can suppose that $0 \leq u_n \leq 1$ and supp $(u_n) \subset conv(D) + B_1$. Now, up to a subsequence, we may suppose that u_n converges in $L^2(\mathbb{R}^d)$ and pointwise almost everywhere to a function $u \in H^1(\mathbb{R}^d)$. The claim follows by the semicontinuity of $\mathcal{F}_\Lambda$.   □

In Propositions 2.15 and 2.16, we will prove that, in the special case when the domains D in (2.3) and (2.4) are balls, the solution is unique and can be computed explicitly.

Proposition 2.15 (Optimal Exterior Domains) *Let the domain D in $\mathbb{R}^d$ be the ball B_r. Then, there is a unique solution u_r of (2.3). Moreover, for every r, there is a radius $R > r$, uniquely determined by r and d, such that u_r is given by*

$$u_r = 1 \quad in \quad B_r, \qquad u_r = 0 \quad in \quad \mathbb{R}^d \setminus B_R \qquad and \qquad u_r = h_r \quad in \quad B_R \setminus B_r,$$

where h_r is a radial harmonic function (as on Fig. 2.2). Precisely, h_r is given by

$$h_r(x) = \frac{|x|^{2-d} - R^{2-d}}{r^{2-d} - R^{2-d}} \quad if \quad d \geq 3, \qquad h_r(x) = \frac{\ln|x| - \ln R}{\ln r - \ln R} \quad if \quad d = 2.$$

Moreover, the radius R and the function u_r satisfy the following properties:

(i) The radius $R = R(r)$ is a continuous function of r such that

$$r < R < r + 1$$

and

$$\lim_{r \to +\infty} |R(r) - (r+1)| = 0.$$

Fig. 2.2 An exterior radial solution

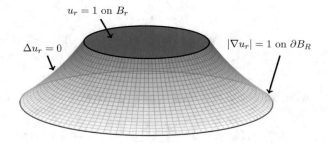

$u_r = 1$ on B_r

$\Delta u_r = 0$

$|\nabla u_r| = 1$ on ∂B_R

(ii) The gradient of h_r is given by

$$|\nabla h_r|(x) = (|x|/R)^{1-d}.$$

Proof We first notice that taking the Schwartz symmetrization u^* of any function u we get that $\mathcal{F}_1(u^*, \mathbb{R}^d) \leq \mathcal{F}_1(u, \mathbb{R}^d)$. Thus, there is a minimizer of $\mathcal{F}_1$ which is a radial function. We first show that there is a unique radial function that minimizes of $\mathcal{F}_1$ in the class of radial functions.

Let $d \geq 3$. For every $0 < r < R$, consider the function

$$u_{r,R}(x) = \begin{cases} 1, & \text{if } |x| \leq r, \\ \dfrac{|x|^{2-d} - R^{2-d}}{r^{2-d} - R^{2-d}}, & \text{if } r < |x| < R, \\ 0, & \text{if } |x| \geq R. \end{cases}$$

Since $u_{r,R}$ is the unique harmonic function in $B_R \setminus B_r$, we get that the minimizer of $\mathcal{F}_1$ among the radial functions is necessarily given by a function of the form $u_{r,R}$. We calculate the energy

$$\mathcal{F}_1(u_{r,R}, \mathbb{R}^d) = \int_{B_R \setminus B_r} |\nabla u_{r,R}|^2 \, dx + |B_R| = \frac{d(d-2)\omega_d}{r^{2-d} - R^{2-d}} + \omega_d R^d.$$

We notice that the function $f(R) := \dfrac{d(d-2)}{r^{2-d} - R^{2-d}} + R^d$ is strictly convex and

$$\lim_{R \to r^+} f(R) = \lim_{R \to +\infty} f(R) = +\infty.$$

Thus, there is a unique radius $R > r$ that minimizes f. We denote this radius by R_*. Notice that, since $f'(R_*) = 0$, we have

$$R_*^{d-1}\left(r^{2-d} - R_*^{2-d}\right) = d - 2. \tag{2.5}$$

Let $d = 2$. For every $0 < r < R$, consider the function

$$u_{r,R}(x) = \begin{cases} 1, & \text{if } |x| \leq r, \\ \dfrac{\ln(R/|x|)}{\ln(R/r)}, & \text{if } r < |x| < R, \\ 0, & \text{if } |x| \geq R. \end{cases}$$

As in the case $d \geq 3$, we calculate the energy

$$\mathcal{F}_1(u_{r,R}, \mathbb{R}^d) = \int_{B_R \setminus B_r} |\nabla u_{r,R}|^2 \, dx + |B_R| = \frac{2\pi}{\ln(R/r)} + \pi R^2.$$

As in the case $d > 2$, there is a unique $R_* > r$ that minimizes the function $R \mapsto \mathcal{F}(u_{r,R})$. Moreover, R_* is such that

$$R_*\big(\ln R_* - \ln r\big) = 1. \tag{2.6}$$

We notice that the claims (i) and (ii) follow by (2.5) and (2.6).

We now prove that the functions u_{r,R_*} are the unique minimizers of $\mathcal{F}_1$ among all admissible functions. Indeed, consider any minimizer u of $\mathcal{F}_1$ and suppose that it is not radial. We notice that the symmetrized function u^* is also a solution. Since it is radial, we get that $u^* = u_{r,R_d^*}$ and in particular $|\{u > 0\}| = |B_{R_*}|$. By Lemma 2.5, the functions $v = u \wedge u^*$ and $V = u \vee u^*$ are also minimizers of $\mathcal{F}$. If u is not radial, then we have $|\{v > 0\}| \neq |B_{R_*}|$ or $|\{V > 0\}| \neq |B_{R_*}|$. On the other hand the symmetrized function v^* and V^* are also solutions and so, we must have $v^* = V^* = u^*$ and in particular $|\{v > 0\}| = |\{V > 0\}| = |B_{R_*}|$, which is in contradiction with the assumption that u is not radially symmetric. $\qquad\square$

Proposition 2.16 (Optimal Interior Domains) *Let the domain D in $\mathbb{R}^d$ be the ball B_R. Then, there is a dimensional constant $C_d > 0$ such that, for every $R > C_d$, there is a unique solution u_R of (2.4). Moreover, u_R is radially symmetric and has the following properties:*

$$u_R = 1 \quad on \quad \partial B_R, \qquad u_R = 0 \quad in \quad B_r \qquad and \qquad u_R = h_R \quad in \quad B_R \setminus B_r, \tag{2.7}$$

where h_R is a radially symmetric harmonic function (see Fig. 2.3). Precisely,

$$h_R(x) = \frac{|x|^{2-d} - r^{2-d}}{R^{2-d} - r^{2-d}} \quad if \quad d \geq 3, \qquad h_R(x) = \frac{\ln|x| - \ln r}{\ln R - \ln r} \quad if \quad d = 2,$$

Fig. 2.3 An interior radial solution

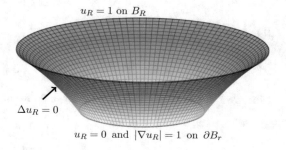

$u_R = 1$ on B_R

$\Delta u_R = 0$

$u_R = 0$ and $|\nabla u_R| = 1$ on ∂B_r

where the radius r depends on R and d and has the following properties:

(i) The radius $r = r(R)$ is a continuous function of R such that

$$\lim_{R \to +\infty} |r(R) - (R-1)| = 0.$$

(ii) The gradient of h_R is given by

$$|\nabla h_R|(x) = (|x|/r)^{1-d}.$$

Proof As in the proof of Lemma 2.15, we start by noticing that for every function u, there is a radially symmetric function u^* with lower energy. In fact, it is sufficient to consider the function $v = 1 - u$ and its Schwartz symmetrization v^*. We define u^* as $u^* := 1 - v^*$ and we notice that

$$\mathcal{F}_1(u^*, B_R) = \int_{B_R} |\nabla u^*|^2 \, dx + |\{u^* > 0\} \cap B_R| = \int_{B_R} |\nabla v^*|^2 \, dx + |\{v^* < 1\} \cap B_R|$$

$$\le \int_{B_R} |\nabla v|^2 \, dx + |\{v < 1\} \cap B_R| = \int_{B_R} |\nabla u|^2 \, dx + |\{u > 0\} \cap B_R| = \mathcal{F}_1(u, B_R).$$

Thus, there exists a radially symmetric minimizer u^* of $\mathcal{F}$. Now, since u^* is harmonic in $\{u^* > 0\}$, it should be of the form $u^* = u_{r,R}$, where $u_{r,R}$ is given by (2.7) for some radius $r < R$. Now, for any $r \in (0, R)$, the energy of $u_{r,R}$ is given by

$$\mathcal{F}_1(u_{r,R}, B_R) = \int_{B_R \setminus B_r} |\nabla u_{r,R}|^2 \, dx + |B_R \setminus B_r| = \frac{d(d-2)\omega_d}{r^{2-d} - R^{2-d}} + \omega_d(R^d - r^d).$$

Consider the function

$$f(r) := \frac{d(d-2)}{r^{2-d} - R^{2-d}} - r^d.$$

It is easy to check that,

$$\lim_{r \to 0} f(r) = 0 \qquad \text{and} \qquad \lim_{r \to R} f(r) = +\infty.$$

Moreover, for R large enough, $f(R/2) < 0$. We now calculate

$$f'(r) = \frac{d(d-2)^2 r^{1-d}}{\left(r^{2-d} - R^{2-d}\right)^2} - dr^{d-1}.$$

Thus, $f'(r) = 0$ if and only if

$$g(r) := (d-2) - r + r^{d-1} R^{2-d} = 0.$$

Now, the equation $g(r) = 0$ has at most two solutions and we have that

$$g(0) = g(R) = d - 2 > 0.$$

On the other hand, for R large enough, we have

$$g(d - 1) < 0 \qquad \text{and} \qquad g(R - 2) < 0.$$

Thus, the equation $g(r) = 0$ has exactly two solutions:

$$r_- \in (0, d - 1) \qquad \text{and} \qquad r_+ \in (R - 2, R).$$

Now, let M_d be the minimum of f in the interval $[0, d - 1]$. For R large enough, we have

$$f(R - 2) = (R - 2)^{d-2} \left(\frac{d(d - 2)}{1 - (1 - 2/R)^{d-2}} - (R - 2)^2 \right) < M_d.$$

Thus, there is a unique $r \in (0, R)$ that minimizes f in $(0, R)$. Moreover, $R - 2 < r < R$. Moreover, the claim (i) follows from the fact that, for every $\varepsilon > 0$, there is $R_\varepsilon > 0$ such that if $R > R_\varepsilon$, then

$$g(R - (1 - \varepsilon)) < 0 \qquad \text{and} \qquad g(R - (1 + \varepsilon)) > 0.$$

This implies that $R - (1 + \varepsilon) \leq r(R) \leq R - (1 - \varepsilon)$, which is precisely (i).

Let now $d = 2$. For every $r \in (0, R)$, consider the function $u_{r,R}$ given by (2.7) for some $r > 0$. We calculate the energy

$$\mathcal{F}_1(u_{r,R}, B_R) = \int_{B_R \setminus B_r} |\nabla u_{r,R}|^2 \, dx + |B_R \setminus B_r| = \frac{2\pi}{\ln(R/r)} + \pi(R^2 - r^2).$$

Next, we define

$$f(r) := \frac{2}{\ln R - \ln r} - r^2,$$

we calculate

$$f'(r) = \frac{2}{r(\ln R - \ln r)^2} - 2r,$$

and we set

$$g(r) := 1 - r(\ln R - \ln r).$$

As above, g can have at most two zeros in the interval $(0, R)$. Moreover, $g(0) = g(R) = 1$ and for R large enough, we have

$$g(1) = 1 - \ln R < 0 \quad \text{and} \quad g(R - 2) = 1 - (R - 2)\ln\left(1 - \frac{2}{R - 2}\right) < 0.$$

Thus, the two zeros of g are in the intervals $(0, 1)$ and $(R - 2, R)$, respectively. Now, for R large enough, we have

$$f(R - 2) = \frac{2}{\ln\left(1 + \frac{2}{R - 2}\right)} - (R - 2)^2 < -1 < f(1).$$

Thus, for large enough R, there is a unique r that minimizes f in $(0, R)$ and $R - 2 < r < R$. The claim (i) follows as in the case $d > 2$. The claim (ii) is immediate and follows from the equation $g(r) = 0$. The uniqueness of the solution now follows as in Lemma 2.15. $\square$

Chapter 3
Lipschitz Continuity of the Minimizers

In this section, we will prove that the local minimizers of $\mathcal{F}_\Lambda$ are Lipschitz continuous. Our main result is the following.

Theorem 3.1 *Let $D \subset \mathbb{R}^d$ be an open set and $u \in H^1_{loc}(D)$. Suppose that u is a local minimizer of $\mathcal{F}_\Lambda$ in D. Then, u is locally Lipschitz continuous in D.*

Theorem 3.1 is a consequence of the more general Theorem 3.2, which can be applied not only to minimizers of $\mathcal{F}_\Lambda$ (we will need this result for the proofs of Theorems 1.2, 1.4 and 1.10), but also to the case of minimizers for the problem with a measure constraint (Theorem 1.9); we notice that we will be able to apply Theorem 3.2 to (1.6) only after proving that an outwards minimality property of the type (3.1) holds at very small scale (see Sect. 11.5).

Theorem 3.2 *Let D be a bounded open set in $\mathbb{R}^d$ and $u \in H^1(D)$ be a non-negative function satisfying the following minimality condition:*

$$\mathcal{F}_\Lambda(u, D) \leq \mathcal{F}_\Lambda(v, D) \quad \text{for every} \quad v \in H^1(D) \quad \text{such that}$$

$$u - v \in H^1_0(D) \quad \text{and} \quad \Omega_u \subset \Omega_v. \tag{3.1}$$

Then, u is locally Lipschitz continuous in D.

The outwards minimality condition appeared recently in [9] in the context of a shape optimization problem, which can be reduced to a free boundary problem for vector-valued functions (see [41]). This property proved to be very useful only in the context of other free boundary and shape optimization problems as, for instance, the ones involving functionals depending on the perimeter of the set (see [21] and [22]). In the case of $\mathcal{F}_\Lambda$ the outwards minimality condition (3.1) can also be expressed in a different way. We give the precise statement in the following lemma.

© The Author(s) 2023
B. Velichkov, *Regularity of the One-phase Free Boundaries*,
Lecture Notes of the Unione Matematica Italiana 28,
https://doi.org/10.1007/978-3-031-13238-4_3

Lemma 3.3 *Let D be a bounded open set in $\mathbb{R}^d$ and $u \in H^1(D)$ be a given non-negative function. Then the following are equivalent:*

(i) u satisfies the minimality condition (3.1);
(ii) u is harmonic in Ω_u in the following sense:

$$\int_D |\nabla u|^2 \, dx \le \int_D |\nabla v|^2 \, dx \quad \text{for every} \quad v \in H^1(D) \quad \text{such that}$$

$$u - v \in H^1(\mathbb{R}^d) \quad \text{and} \quad u - v = 0 \quad \text{a.e. in} \quad \mathbb{R}^d \setminus \Omega_u,$$
$$(3.2)$$

and satisfies the minimality condition

$$\mathcal{F}_\Lambda(u, D) \le \mathcal{F}_\Lambda(v, D) \quad \text{for every} \quad v \in H^1(D) \quad \text{such that}$$

$$u - v \in H^1_0(D) \quad \text{and} \quad u \le v \quad \text{in} \quad D.$$
$$(3.3)$$

Remark 3.4 (On the Sign of the Test Functions in (3.1), (3.2) and (3.3)) Since u is non-negative in D, we may suppose that the test functions v in (3.1), (3.2) and (3.3) are all non-negative.

Proof of Lemma 3.3 The fact that (3.1) implies (3.2) and (3.3) is trivial. Suppose now that u satisfies both (3.2) and (3.3) and let $v \in H^1(D)$ be a non-negative function such that $u - v \in H^1_0(D)$ and $\Omega_u \subset \Omega_v$. Then consider the test functions $u \wedge v$ and $u \vee v$. Since $u \wedge v = 0$ outside Ω_u, by (3.2), we have that

$$\int_D |\nabla u|^2 \, dx \le \int_D |\nabla (u \wedge v)|^2 \, dx.$$

On the other hand, since $u \vee v \ge u$, (3.3) implies that

$$\int_D |\nabla u|^2 \, dx + \Lambda |\Omega_u| \le \int_D |\nabla (u \vee v)|^2 \, dx + \Lambda |\Omega_{u \vee v}|.$$

Summing up the two inequalities, we get

$$2 \int_D |\nabla u|^2 \, dx + \Lambda |\Omega_u| \le \int_D |\nabla (u \wedge v)|^2 \, dx + \int_D |\nabla (u \vee v)|^2 \, dx + \Lambda |\Omega_{u \vee v}|$$

$$= \int_D |\nabla u|^2 \, dx + \int_D |\nabla v|^2 \, dx + \Lambda |\Omega_v|,$$

which is precisely (3.1).   □

We will give three different proofs of Theorem 3.2, but in each one of them, the conclusion (the Lipschitz continuity of u) will be a consequence of the following estimate on the growth of the function u at the free boundary

$$\fint_{\partial B_r(x_0)} u\, d\mathcal{H}^{d-1} \leq C\, r \quad \text{for every} \quad x_0 \in \partial \Omega_u \quad \text{and every} \quad 0 < r < r_0,$$

$$(3.4)$$

where $r_0 > 0$ and $C > 0$ are universal constants depending on the distance to the boundary ∂D. We give the precise statement in the following lemma.

Lemma 3.5 *Suppose that $u \in H^1(D)$ is a non-negative function such that:*

- *u is harmonic in the interior of the set $\Omega_u := \{u > 0\}$;*
- *u satisfies the inequality (3.4) with constants C and r_0 uniformly in D.*

Then the set Ω_u is open and the function u is locally Lipschitz continuous in D. Precisely, the gradient of u can be estimated as

$$\|\nabla u\|_{L^\infty(D_\delta)} \leq C_d \left(C + \frac{\|u\|_{L^1(D_{\delta/2})}}{\delta^{d+1}} \right) \quad \text{for every} \quad 0 < \delta < r_0,$$

where C_d is a dimensional constant and, for $r > 0$, we use the notation

$$D_r := \left\{ x \in D \ : \ \operatorname{dist}(x, \partial D) > r \right\}.$$

Proof Suppose that $x_0 \in D \cap \partial \Omega_u$. Passing to the limit as $r \to 0$ the estimate (3.4) we obtain that $u(x_0) = 0$. Thus $\Omega_u \cap \partial \Omega_u = \emptyset$ and so Ω_u is open.

Let now $x_0 \in D_\delta$. We consider two cases.

- If $\operatorname{dist}(x_0, \partial \Omega_u) \geq \delta/4$, then u is harmonic in the ball $B_{\delta/4}(x_0)$ and so, by the gradient estimate (see for example [30]) we have

$$|\nabla u(x_0)| \leq \frac{C_d}{\delta^{d+1}} \int_{B_\delta(x_0)} u\, dx,$$

where C_d is a dimensional constant.

- If $\operatorname{dist}(x_0, \partial \Omega_u) < \delta/4$, then we suppose that the distance to the free boundary is realized by some $y_0 \in \partial \Omega_u$ and we set

$$r = \operatorname{dist}(x_0, \partial \Omega_u) = |x_0 - y_0|.$$

Since u is harmonic in $B_r(x_0)$, we can again apply the gradient estimate obtaining

$$|\nabla u(x_0)| \leq \frac{C_d}{r^{d+1}} \int_{B_r(x_0)} u\, dx \leq \frac{C_d}{r^{d+1}} \int_{B_{2r}(y_0)} u\, dx \leq C_d\, C,$$

where the second inequality follows by the positivity of u and the inclusion $B_r(x_0) \subset B_{2r}(y_0)$. The last inequality is simply a consequence of (3.4) and the fact that

$$\int_{B_{2r}(y_0)} u\, dx = \int_0^{2r} ds \int_{\partial B_s(y_0)} u\, d\mathcal{H}^{d-1}.$$

$\square$

Remark 3.6 (An Alternative Statement of (3.4)) We notice that (3.4) is a consequence of the following inequality

$$\fint_{\partial B_r(x_0)} u\, d\mathcal{H}^{d-1} \leq C\, r \quad \text{for every} \quad x_0 \in \{u = 0\} \quad \text{and every} \quad 0 < r < r_0.$$

(3.5)

This is trivial if we knew a priori that u is continuous, but is true also in general. Indeed, by Lemma 2.9, we have that

$$\partial\Omega_u = \Big\{x_0 \in D \; : \; 0 < |\Omega_u \cap B_r(x_0)| < |B_r| \quad \text{for every} \quad r > 0\Big\}.$$

Thus, every point $x_0 \in \partial\Omega_u$ can be obtained as limit of points $x_n \in \{u = 0\}$, for which the estimate (3.5) does hold. The claim follows by the continuity of the function

$$x \mapsto \fint_{\partial B_r(x)} u\, d\mathcal{H}^{d-1},$$

for every fixed $r > 0$, which is due to the fact that $u \in H^1(D)$.

The rest of this section is dedicated to the proof of (3.4) in the hypotheses of Theorem 3.2. In the next three subsections we will give three different proofs of this fact.

- **Section 3.1.** *The Alt-Caffarelli proof of the Lipschitz continuity.*

    In this section we present the original proof proposed by Alt and Caffarelli (see [3]), which we divide in two steps (Lemmas 3.7 and 3.8). This entire section comes directly from [51] and we report it here for the sake of completeness.
- **Section 3.2.** *The Laplacian estimate.*

    In this section we give a proof, which is inspired from the proof of the Lipschitz continuity of the solution to the two-phase problem, which was given by Alt, Caffarelli and Friedman in [4]. In our case there is only one phase (that is, the solution u is positive), so we do not make use of the two-phase monotonicity formula of Alt-Caffarelli-Friedman, which significantly simplifies the proof. This approach can be used also in other situations, for instance, for functionals involving elliptic operators (in divergence form) with non-constant coefficients (see [46]).

- **Section 3.3.** *The Danielli-Petrosyan approach.*

This last subsection is dedicated to the method proposed by Danielli and Petrosyan in [18] in the context of non-linear operators. It consists of two steps. The first one is to show that u is Hölder continuous. This part of the argument is very general and is based on classical regularity estimates for (almost-)minimizers of variational problems. In the second step of the proof, the Lipschitz continuity is obtained by absurd and the result of the first step is used to assure the convergence of the sequence of minimizers produced by contradiction. This type of argument (proving a weaker estimate and then obtaining the main result by contradiction) will be used also in Chap. 8, this time to obtain the regularity of the free boundary.

3.1 The Alt-Caffarelli's Proof of the Lipschitz Continuity

This subsection contains the original argument proposed by Alt and Caffarelli in [3]. The main steps of the proof are the following:

- Comparing the energy $\mathcal{F}_\Lambda(u, B_r(x_0))$ of u in the ball $B_r(x_0)$ with the one of the harmonic extension h of u in $B_r(x_0)$ we get

$$\int_{B_r(x_0)} |\nabla(u - h)|^2 \, dx \leq \Lambda \big| \{u = 0\} \cap B_r(x_0) \big|.$$

- It is now sufficient to estimate from below the right-hand side of the above inequality. In Lemma 3.7 we will prove that

$$\frac{1}{r^2} \big| \{u = 0\} \cap B_r(x_0) \big| \left(\fint_{\partial B_r(x_0)} u \, d\mathcal{H}^{d-1} \right)^2 \leq C_d \int_{B_r(x_0)} |\nabla(u - h)|^2 \, dx.$$

- If $x_0 \in \Omega_u$, then $\big| \{u = 0\} \cap B_r(x_0) \big| \neq 0$. Combining the two inequalities we get

$$\frac{1}{r} \fint_{\partial B_r(x_0)} u \, d\mathcal{H}^{d-1} \leq \sqrt{C_d \Lambda}.$$

We now give the details of the proof sketched above. The key ingredient is the following trace-type inequality (Lemma 3.7), which is implicitly contained in the proof of the Lipschitz continuity given in [3] (and can also be found in [51]) and is an interesting result by itself.

Lemma 3.7 *For every $u \in H^1(B_r)$ we have the following estimate:*

$$\frac{1}{r^2} |\{u = 0\} \cap B_r| \left(\fint_{\partial B_r} u \, d\mathcal{H}^{d-1} \right)^2 \leq C_d \int_{B_r} |\nabla(u - h)|^2 \, dx, \qquad (3.6)$$

where:

- C_d *is a constant that depends only on the dimension d;*
- h *is the harmonic replacement of u in B_r, that is, the harmonic function in B_r such that $u = h$ on ∂B_r.*

Proof We report here the proof for the sake of completeness, and refer the reader to [3, Lemma 3.2]. We note that it is sufficient to prove the result in the case $u \geq 0$. Let $v \in H^1(B_r)$ be the solution of the problem

$$\min \left\{ \int_{B_r} |\nabla v|^2 \, dx : \ u - v \in H_0^1(B_r), \ v \geq u \right\}.$$

Notice that v is super-harmonic on B_r and harmonic on the set $\{v > u\}$.

For each $|z| \leq 1/2$, we consider the functions u_z and v_z defined on B_r as

$$u_z(x) := u\big((r - |x|)z + x\big) \qquad \text{and} \qquad v_z(x) := v\big((r - |x|)z + x\big).$$

Note that both u_z and v_z still belong to $H^1(B_r)$ and that their gradients are controlled from above and below by the gradients of u and v. We call S_z the set of all $|\xi| = 1$ such that the set $\left\{ \rho : \frac{r}{8} \leq \rho \leq r, \ u_z(\rho \xi) = 0 \right\}$ is not empty. For $\xi \in S_z$ we define

$$r_\xi = \inf \left\{ \rho : \frac{r}{8} \leq \rho \leq r, \ u_z(\rho \xi) = 0 \right\}.$$

For almost all $\xi \in S^{d-1}$ (and then for almost all $\xi \in S_z$), the functions $\rho \mapsto \nabla u_z(\rho \xi)$ and $\rho \mapsto \nabla v_z(\rho \xi)$ are square integrable. For those ξ, one can suppose that the equation

$$\big((u_z(\rho_2 \xi) - v_z(\rho_2 \xi)) - (u_z(\rho_1 \xi) - v_z(\rho_1 \xi))\big) = \int_{\rho_1}^{\rho_2} \xi \cdot \nabla\big(u_z(\rho \xi) - v_z(\rho \xi)\big) \, d\rho,$$

holds for all $\rho_1, \rho_2 \in [0, r]$. Moreover, we have the estimate

$$v_z(r_\xi \xi) = \int_{r_\xi}^{r} \xi \cdot \nabla(v_z - u_z)(\rho \xi) \, d\rho \leq \sqrt{r - r_\xi} \left(\int_{r_\xi}^{r} |\nabla(v_z - u_z)(\rho \xi)|^2 \, d\rho \right)^{1/2}.$$

Since v is superharmonic we have that, by the Poisson's integral formula,

$$v(x) \geq h(x) = \frac{r^2 - |x|^2}{d\omega_d r} \int_{\partial B_r} \frac{u(y)}{|x-y|^d} \, d\mathcal{H}^{d-1}(y) \geq c_d \frac{r - |x|}{r} \fint_{\partial B_r} u \, d\mathcal{H}^{d-1},$$

where h is the harmonic function such that $h = u(= v)$ on ∂B_r. Taking

$$x = (r - r_\xi)z + r_\xi \xi,$$

we have

$$v_z(r_\xi \xi) = v\big((r - r_\xi)z + r_\xi \xi\big) \geq \frac{c_d}{2} \frac{r - r_\xi}{r} \fint_{\partial B_r} u \, d\mathcal{H}^{d-1} = \frac{c_d}{2} \frac{r - r_\xi}{r} \fint_{\partial B_r} u_z \, d\mathcal{H}^{d-1}.$$

Combining the two inequalities, we have

$$\frac{r - r_\xi}{r^2} \left(\fint_{\partial B_r} u \, d\mathcal{H}^{d-1} \right)^2 \leq C_d \int_{r_\xi}^{r} |\nabla(v_z - u_z)|^2 (\rho \xi) \, d\rho.$$

Integrating over $\xi \in S_z \subset S^{d-1}$, we obtain the inequality

$$\left(\int_{S_z} \frac{r - r_\xi}{r^2} \, d\xi \right) \left(\fint_{\partial B_r} u \, d\mathcal{H}^{d-1} \right)^2 \leq C_d \int_{\partial B_1} \int_{r_\xi}^{r} |\nabla(v_z - u_z)(\rho \xi)|^2 \, d\rho \, d\xi,$$

and, by the estimate that $r/8 \leq r_\xi \leq r$, we have

$$\frac{1}{r^2} |\{u = 0\} \cap B_r \setminus B_{r/4}(rz)| \left(\fint_{\partial B_r} u \, d\mathcal{H}^{d-1} \right)^2 \leq C_d \int_{B_r} |\nabla(v_z - u_z)|^2 \, dx$$

$$\leq C_d \int_{B_r} |\nabla(v - u)|^2 \, dx.$$

Integrating over z, we obtain (3.6). $\square$

Lemma 3.8 *Suppose that $u \in H^1_{loc}(D)$ be a local minimizer of $\mathcal{F}_\Lambda$ in the open set $D \subset \mathbb{R}^d$. Then for every ball $\overline{B}_r(x_0) \subset D$ we have*

$$|\{u = 0\} \cap B_r(x_0)| \left(\sqrt{C_d \Lambda} - \frac{1}{r} \fint_{\partial B_r(x_0)} u \, d\mathcal{H}^{d-1} \right) \geq 0.$$

In particular, if $x_0 \in \partial \Omega_u$, then

$$\fint_{\partial B_r(x_0)} u \, d\mathcal{H}^{d-1} \leq C_d \sqrt{\Lambda}\, r.$$

Proof Suppose that $x_0 = 0$. Let $h \in H^1(B_r)$ be the harmonic function in B_r such that $h = u$ on ∂B_r. By the optimality of u we get

$$\int_{B_r} |\nabla u|^2 \, dx + \Lambda |\{u > 0\} \cap B_r| \le \int_{B_r} |\nabla h|^2 \, dx + \Lambda |B_r|.$$

Now using (3.6) and the fact that

$$\int_{B_r} |\nabla (u - h)|^2 \, dx = \int_{B_r} \left(|\nabla u|^2 - |\nabla h|^2 \right) dx \le \Lambda |\{u = 0\} \cap B_r|,$$

we get

$$|\{u = 0\} \cap B_r| \left(\sqrt{C_d \Lambda} - \frac{1}{r} \fint_{\partial B_r} u \, d\mathcal{H}^{d-1} \right) \left(\sqrt{C_d \Lambda} + \frac{1}{r} \fint_{\partial B_r} u \, d\mathcal{H}^{d-1} \right) \ge 0,$$

which gives the claim. $\square$

3.2 The Laplacian Estimate

In this section, we propose a different approach to the Lipschitz continuity of u. The method comes from the two-phase free boundary theory and, in particular, from the work of Alt-Caffarelli-Friedman [4] and Briançon-Hayouni-Pierre [7]. This argument was also adapted to the vectorial case in [41] and to a one-phase shape optimization problem in [46]. The proof consists of two steps:

- For every local minimizer u of $\mathcal{F}_1$ we have that Δu is a positive measure. In Lemma 3.9, we prove that the optimality of u implies the estimate

$$\Delta u(B_r) \le C \, r^{d-1}.$$

- In Lemma 3.10, we show that the Laplacian estimate and the classical identity

$$\frac{d}{dr} \fint_{\partial B_r} u \, d\mathcal{H}^{d-1} = \frac{\Delta u(B_r)}{d \, \omega_d \, r^{d-1}},$$

imply that

$$\fint_{\partial B_r} u \, d\mathcal{H}^{d-1} \le Cr,$$

which gives the Lipschitz continuity of u by Proposition 3.5.

Lemma 3.9 (The Laplacian Estimate) *Suppose that u is a local minimizer of $\mathcal{F}_1$ in D. Then, for every ball $B_r(x_0)$ such that $B_{2r}(x_0) \subset D$ we have*

$$\Delta u(B_r(x_0)) \le C\, r^{d-1}.$$

Proof Without loss of generality we can assume that $x_0 = 0$. We now notice that by Lemma 2.6 the distributional Laplacian

$$\Delta u(\varphi) := -\int_D \nabla u \cdot \nabla \varphi\, dx \quad \text{for every} \quad \varphi \in C_c^1(D),$$

is a positive Radon measure. We first prove that

$$\Delta u(\varphi) \le C_d\, r^{d/2} \|\nabla \varphi\|_{L^2(B_r)} \quad \text{for every} \quad \varphi \in C_c^\infty(B_r) \quad \text{and every} \quad B_r \subset D. \tag{3.7}$$

Indeed, for every $\psi \in C_c^\infty(B_r)$, the optimality of u gives

$$\int_{B_r} |\nabla u|^2\, dx \le \int_{B_r} |\nabla u|^2\, dx + |\{u > 0\} \cap B_r| \le \int_{B_r} |\nabla(u + \psi)|^2\, dx + |B_r|.$$

Developing the gradient on the right-hand side, we get

$$-\int_{B_r} \nabla u \cdot \nabla \psi\, dx \le \frac{1}{2}\left(\int_{B_r} |\nabla \psi|^2\, dx + \omega_d\, r^d\right).$$

Setting $\psi = r^{d/2}\, \|\nabla \varphi\|_{L^2(B_r)}^{-1}\, \varphi$, we get

$$-\int_{B_r} \nabla u \cdot \nabla \varphi\, dx \le \frac{1 + \omega_d}{2}\, r^{d/2} \|\nabla \varphi\|_{L^2(B_r)},$$

which is precisely (3.7) with $C_d = \dfrac{1 + \omega_d}{2}$.

Let now $\varphi \in C_c^\infty(B_{2r})$ be such that

$$\varphi \ge 0 \text{ on } B_{2r}, \qquad \varphi = 1 \text{ on } B_r, \qquad \text{and} \qquad \|\nabla \varphi\|_{L^\infty(B_{2r})} \le \frac{2}{r}.$$

Thus, $\varphi \ge \mathbb{1}_{B_r}$ and by the positivity of Δu we have

$$\Delta u(B_r) \le \Delta u(\varphi) \le C_d\, (2r)^{d/2} \|\nabla \varphi\|_{L^2(B_{2r})} \le C r^{d-1}.$$

$\square$

Now the estimate (3.4) follows by the following lemma.

Lemma 3.10 *Suppose that $u \in H^1(B_R)$ is a non-negative sub-harmonic function in the ball $B_R \subset \mathbb{R}^d$ such that $u(0) = 0$. Suppose that there is a constant $C > 0$ such that*

$$\Delta u(B_r) \leq C r^{d-1} \quad \text{for every} \quad 0 < r < R. \tag{3.8}$$

Then we have

$$\fint_{\partial B_r} u \, d\mathcal{H}^{d-1} \leq \frac{C}{d \omega_d} r \quad \text{for every} \quad 0 < r < R. \tag{3.9}$$

Proof We first notice that for every smooth u_ε we have

$$\frac{d}{dr} \fint_{\partial B_r} u_\varepsilon \, d\mathcal{H}^{d-1} = \fint_{\partial B_r} \frac{\partial u_\varepsilon}{\partial n} \, d\mathcal{H}^{d-1} = \frac{1}{d \, \omega_d \, r^{d-1}} \int_{B_r} \Delta u_\varepsilon(x) \, dx.$$

Integrating in r and passing to the limit as $\varepsilon \to 0$ we get

$$\fint_{\partial B_r} u \, d\mathcal{H}^{d-1} \leq \int_0^r \frac{\Delta u(B_r)}{d \, \omega_d \, r^{d-1}} \, dr.$$

Now, using (3.8) we get (3.9). $\square$

3.3   The Danielli-Petrosyan Approach

Finally, in the last section dedicated to the Lipschitz continuity of the minimizers, we present another proof, which is due to Danielli and Petrosyan and was originally carried out in the framework of the p-laplacian (see [18]). In fact, this proof is very close in spirit to the one of the regularity of the free boundary that we will present in Chap. 8. It consists of two steps. The first one is to prove that the local minimizers are Hölder continuous and to find a uniform estimate on their $C^{0,\alpha}$ norm (see Lemma 3.11, Lemma 3.12 and Proposition 3.13). Then, the Lipschitz continuity (see Proposition 3.15) follows by a contradiction argument, in which the compactness is a consequence of the aforementioned uniform $C^{0,\alpha}$ estimate.

Lemma 3.11 *Suppose that $\Omega \subset \mathbb{R}^d$ is a bounded open set and that the function $u \in H^1(\Omega) \cap L^\infty(\Omega)$ is such that:*

(a) u is non-negative and subharmonic in Ω;
(b) u satisfies the minimality condition (3.3) for some constant $\Lambda > 0$.

Then, setting

$$\varepsilon = \frac{2}{d}, \qquad \alpha = \frac{2}{2+d} \qquad and \qquad C = 2^{d+3}|B_1|\left(\Lambda + \|u\|_{L^\infty(\Omega)}^2\right),$$

the following inequality does hold:

$$\int_{B_\rho(x_0)} |\nabla u|^2\, dx \le C\rho^{d-2(1-\alpha)} \quad \text{for every } B_\rho(x_0) \subset \Omega \text{ with } \rho \le 2^{-\frac{d+2}{2}}.$$

Proof Let $r = \rho^{\frac{1}{1+\varepsilon}}$. Thus we have $B_r(x_0) \subset \Omega$. Without loss of generality we can assume that $x_0 = 0$. Let h be the harmonic extension of u in the ball B_r. Then, $u \le h$ and, by the optimality of u, we get

$$\int_{B_r} |\nabla(u-h)|^2\, dx = \int_{B_r} |\nabla u|^2\, dx - \int_{B_r} |\nabla h|^2\, dx \le \Lambda|B_r|.$$

Thus, we can estimate the gradient of u as follows

$$\int_{B_{r^{1+\varepsilon}}} |\nabla u|^2\, dx \le 2\int_{B_{r^{1+\varepsilon}}} |\nabla(u-h)|^2\, dx + 2\int_{B_{r^{1+\varepsilon}}} |\nabla h|^2\, dx$$

$$\le 2\int_{B_r} |\nabla(u-h)|^2\, dx + 2\frac{|B_{r^{1+\varepsilon}}|}{|B_{r/2}|}\int_{B_{r/2}} |\nabla h|^2\, dx$$

$$\le 2\Lambda|B_r| + 2^{d+1}r^{\varepsilon d}\int_{B_{r/2}} |\nabla h|^2\, dx,$$

where the second inequality follows by the fact that $|\nabla h|^2$ is subharmonic in B_r and the inequality $r^\varepsilon \le 1/2$. Now, we use the Caccioppoli inequality

$$\int_{B_{r/2}} |\nabla h|^2\, dx \le \int_{B_r} |\nabla(h\varphi)|^2\, dx = \int_{B_r} |\nabla\varphi|^2 h^2\, dx \le \|\nabla\varphi\|_{L^\infty}^2 \int_{B_r} h^2\, dx \le \frac{4|B_r|M^2}{r^2},$$

where $M = \|u\|_{L^\infty(D)} \ge \|h\|_{L^\infty(B_r)}$ and φ is given by

$$\varphi(x) = 0 \text{ if } |x| \ge r, \qquad \varphi(x) = 1 \text{ if } |x| \le \frac{r}{2}, \qquad \varphi(x) = \frac{2}{r}(r - |x|) \text{ if } \frac{r}{2} < |x| < r.$$

Since $\rho = r^{1+\varepsilon}$ and $\varepsilon = 2/d$ we obtain

$$\int_{B_\rho} |\nabla u|^2\, dx \le 2\Lambda|B_1|\rho^{\frac{d}{1+\varepsilon}} + 2^{d+3}|B_1|M^2\rho^{d-\frac{2}{1+\varepsilon}} \le 2^{d+3}|B_1|(\Lambda + M^2)\rho^{\frac{d}{1+\varepsilon}}$$

which gives the claim. $\qquad\qquad\qquad\qquad\qquad\qquad\qquad\qquad\qquad\qquad\qquad\qquad \square$

Lemma 3.12 (Morrey) *Suppose that $\Omega \subset \mathbb{R}^d$, $u \in H^1(B_R)$ and that there are constants $C > 0$ and $\alpha \in (0,1)$ such that*

$$\fint_{B_r(x_0)} |\nabla u|^2 \, dx \le Cr^{2(\alpha-1)} \quad \text{for every} \quad x_0 \in B_{R/8} \quad \text{and every} \quad r \le R/2.$$

Then $u \in C^{0,\alpha}(B_{R/8})$ and

$$|u(x) - u(y)| \le \sqrt{C}\left(2^d + \frac{2}{\alpha}\right)|x - y|^\alpha \quad \text{for every} \quad x, y \in B_{R/8}.$$

Proof Suppose that $x, y \in B_{R/8}$ and let $r = |x - y|$.

$$\left|\fint_{B_r(x)} u - \fint_{B_r(y)} u\right| = \left|\fint_{B_r} [u(x+z) - u(y+z)]\,dz\right|$$

$$= \left|\fint_{B_r} dz \int_0^1 (y-x)\cdot\nabla u(x(1-t)+ty+z)\,dt\right|$$

$$\le |x-y|\fint_{B_r} dz \int_0^1 |\nabla u(x(1-t)+ty+z)|\,dt$$

$$= |x-y|\int_0^1 dt \fint_{B_r} |\nabla u(x(1-t)+ty+z)|\,dz$$

$$\le |x-y|\int_0^1 dt \fint_{B_{2r}(x)} |\nabla u| = r\fint_{B_{2r}(x)} |\nabla u|$$

$$\le r|B_{2r}|\left(\fint_{B_{2r}(x)} |\nabla u|^2\right)^{1/2} \le 2^d\sqrt{C}|B_r|r^\alpha.$$

Let now $x_0 \in B_{R/8}$ be fixed. Assume for simplicity that $x_0 = 0$. Then we have

$$\fint_{B_r} u - \fint_{B_s} u = \fint_{B_1} [u(rx) - u(sx)]\,dx = \fint_{B_1} dx \int_s^r x\cdot\nabla u(tx)\,dt$$

$$\le \fint_{B_1} dx \int_s^r |\nabla u(tx)|\,dt = \int_s^r dt \fint_{B_1} |\nabla u(tx)|\,dx = \int_s^r dt \fint_{B_t} |\nabla u|\,dx$$

$$\le \int_s^r dt \left(\fint_{B_t} |\nabla u|^2\,dx\right)^{1/2} \le \int_s^r \sqrt{C}t^{\alpha-1}\,dt \le \frac{\sqrt{C}}{\alpha}r^\alpha,$$

which concludes the proof. $\square$

The following proposition is a direct consequence of Lemmas 3.11 and 3.12.

Proposition 3.13 (A Uniform Hölder Estimate) *Suppose that the non-negative function $u \in H^1(B_1) \cap L^\infty(B_1)$ satisfies the minimality condition (3.1) in the set $D = B_1$. Then, there is a dimensional constant C_d and a universal numerical constant $\rho > 0$ (one may take $\rho = 1/8$) such that*

$$\int_{B_\rho} |\nabla u|^2 \, dx \leq C_d \left(\Lambda + \|u\|_{L^\infty(B_1)}^2 \right),$$

and

$$|u(x) - u(y)| \leq C_d \left(\Lambda + \|u\|_{L^\infty(B_1)}^2 \right)^{\frac{1}{2}} |x - y|^{\frac{2}{2+d}} \qquad \text{for every} \qquad x, y \in B_\rho.$$

We are now in position to prove the Lipschitz continuity of u. The idea is to argue by contradiction. In fact, suppose that there is a sequence of functions u_k that minimize the functional $\mathcal{F}_\Lambda$ in B_1 and are such that $u_k(0) = 0$ and $m_k := \|u_k\|_{L^\infty(B_{1/2})} \to +\infty$. Then, the functions $v_k = m_k^{-1} u_k$ minimize $\mathcal{F}_{\Lambda/m_k}$ and are such that $v_k(0) = 0$ and $\|v_k\|_{L^\infty(B_{1/2})} = 1$. Now, if v_k converges to some v_∞ weakly in $H^1(B_{1/2})$, then v_∞ is harmonic in $B_{1/2}$. Moreover, if the convergence is also uniform, then $v_\infty(0) = 0$, $v_\infty \geq 0$ in $B_{1/2}$ and $\|v_\infty\|_{L^\infty(B_{1/2})} = 1$, which is impossible. Now, there are two main difficulties that we will have to deal with.

- The first one is the compactness of the sequence v_k. Notice that the L^∞ bound of v_k in $B_{1/2}$ only assures the uniform $C^{0,\alpha}$ bound strictly inside $B_{1/2}$. On the other hand if v_k converges uniformly to zero inside $B_{1/2}$ there wouldn't be any contradiction at the limit. Thus, we will need an Harnack-type inequality in order to assure that v_k remains bounded from below also inside $B_{1/2}$. We will solve this issue in the proof of Proposition 3.15.
- The second issue is the harmonicity of v_∞, which will be a consequence of Lemma 3.14 below.

Lemma 3.14 (Convergence of Local Minimizers) *Let $B_R \subset \mathbb{R}^d$ and u_n be a sequence of non-negative functions in $H^1(B_R)$ such that:*

(a) every u_n satisfies the quasi-minimality condition

$$\mathcal{F}_0(u_n, B_R) \leq \mathcal{F}_0(u_n + \varphi, B_R) + \varepsilon_n$$

$$\text{for every} \quad \varphi \in H_0^1(B_r) \quad \text{and every} \quad r < R, \tag{3.10}$$

where ε_n is a vanishing sequence of positive constants.

(b) the sequence u_n is uniformly bounded in $H^1(B_R)$, that is, for some constant $C > 0$,

$$\|u_n\|^2_{H^1(B_R)} = \mathcal{F}_0(u_n, B_R) + \int_{B_R} u_n^2 \, dx \le C \qquad \text{for every} \qquad n \ge 1.$$

Then, there a non-negative $u_\infty \in H^1(B_R)$ such that, up to a subsequence, we have:

(i) u_n converges to u_∞ strongly in $H^1(B_r)$, for every $0 < r < R$;
(ii) u_∞ is harmonic in B_R.

Proof Up to extracting a subsequence, we can suppose that the sequence u_n converges to a function $u_\infty \in H^1(B_R)$ weakly in $H^1(B_R)$, strongly in $L^2(B_R)$ and a.e. in B_R. The weak H^1-convergence implies that for every $0 < r \le R$

$$\|\nabla u_\infty\|_{L^2(B_r)} \le \liminf_{n \to \infty} \|\nabla u_n\|_{L^2(B_r)}, \tag{3.11}$$

with an equality, if and only if, (up to a subsequence) the convergence is strong in B_r. Up to extracting a subsequence we may assume that the limits in the right-hand side of (3.11) do exist. In order to prove (i), we will show that, for fixed $0 < r < R$, we have

$$\|\nabla u_\infty\|_{L^2(B_r)} = \lim_{n \to \infty} \|\nabla u_n\|_{L^2(B_r)}. \tag{3.12}$$

Let $\eta : B_R \to \mathbb{R}$ be a function such that

$$\eta \in C^\infty(B_R), \quad 0 \le \eta \le 1 \quad \text{in} \quad B_R, \quad \eta = 1 \quad \text{on} \quad \partial B_R, \quad \eta = 0 \quad \text{on} \quad B_r. \tag{3.13}$$

Consider the test function $\tilde{u}_n = \eta u_n + (1 - \eta)u_\infty$. Since u_n satisfies the (quasi-)minimality condition (3.10), we have

$$\int_{B_R} |\nabla u_n|^2 \, dx \le \int_{B_R} |\nabla \tilde{u}_n|^2 \, dx + \varepsilon_n.$$

Next, since

$$|\nabla \tilde{u}_n|^2 = \left|\nabla(\eta u_n + (1 - \eta)u_\infty)\right|^2 = \left|(u_n - u_\infty)\nabla\eta + \eta\nabla u_n + (1 - \eta)\nabla u_\infty\right|^2,$$

and since $u_n \to u_\infty$ strongly in $L^2(B_R)$, we have

$$\limsup_{n \to \infty} \int_{B_R} \left(|\nabla \tilde{u}_n|^2 - |\nabla u_n|^2\right) dx$$

$$= \limsup_{n \to \infty} \int_{B_R} \left(\left|(u_n - u_\infty)\nabla\eta + \eta\nabla u_n + (1 - \eta)\nabla u_\infty\right|^2 - |\nabla u_n|^2\right) dx$$

$$= \limsup_{n \to \infty} \int_{B_R} \left((\eta^2 - 1)|\nabla u_n|^2 + 2\eta(1 - \eta)\nabla u_n \cdot \nabla u_\infty + (1 - \eta)^2 |\nabla u_\infty|^2 \right) dx$$

$$= \limsup_{n \to \infty} \int_{B_R} (1 - \eta^2)\left(|\nabla u_\infty|^2 - |\nabla u_n|^2 \right) dx$$

$$\leq \limsup_{n \to \infty} \int_{\{\eta=0\}} \left(|\nabla u_\infty|^2 - |\nabla u_n|^2 \right) dx + \int_{B_R \setminus \{\eta=0\}} |\nabla u_\infty|^2 \, dx. \qquad (3.14)$$

By the weak H^1 convergence of u_n to u_∞ on the set $\{\eta = 0\} \setminus B_r$, we have

$$\int_{\{\eta=0\} \setminus B_r} |\nabla u_\infty|^2 \, dx \leq \liminf_{n \to \infty} \int_{\{\eta=0\} \setminus B_r} |\nabla u_n|^2 \, dx \,,$$

which implies

$$\limsup_{n \to \infty} \int_{\{\eta=0\}} \left(|\nabla u_\infty|^2 - |\nabla u_n|^2 \right) dx \leq \limsup_{n \to \infty} \int_{B_r} \left(|\nabla u_\infty|^2 - |\nabla u_n|^2 \right) dx$$

$$+ \limsup_{n \to \infty} \int_{\{\eta=0\} \setminus B_r} \left(|\nabla u_\infty|^2 - |\nabla u_n|^2 \right) dx$$

$$\leq \limsup_{n \to \infty} \int_{B_r} \left(|\nabla u_\infty|^2 - |\nabla u_n|^2 \right) dx. \qquad (3.15)$$

On the other hand, the optimality of u_n gives

$$0 = \lim_{n \to \infty} \varepsilon_n \leq \limsup_{n \to \infty} \int_{B_R} \left(|\nabla \tilde{u}_n|^2 - |\nabla u_n|^2 \right) dx \,. \qquad (3.16)$$

Finally, (3.14), (3.15), and (3.16) give

$$0 \leq \limsup_{n \to \infty} \int_{B_r} \left(|\nabla u_\infty|^2 - |\nabla u_n|^2 \right) dx + \int_{\{\eta>0\}} |\nabla u_\infty|^2 \, dx \,,$$

which can be re-written as

$$\liminf_{n \to \infty} \int_{B_r} |\nabla u_n|^2 \, dx \leq \int_{B_r} |\nabla u_\infty|^2 \, dx + \int_{\{\eta>0\}} |\nabla u_\infty|^2 \, dx.$$

Now, since η is arbitrary, we finally obtain

$$\liminf_{n \to \infty} \int_{B_r} |\nabla u_n|^2 \, dx \leq \int_{B_r} |\nabla u_\infty|^2 \, dx,$$

which concludes the proof of (i).

We now prove (ii). Let $0 < r < R$ and $\varphi \in H_0^1(B_r)$. It is enough to show that

$$\int_{B_R} |\nabla u_\infty|^2 \, dx \le \int_{B_R} |\nabla(u_\infty + \varphi)|^2 \, dx. \tag{3.17}$$

Let $\eta : B_R \to \mathbb{R}$ be a function that satisfies (3.13) and is such that the set $\mathcal{N} := \{\eta < 1\}$ is a ball strictly contained in B_R. Notice that

$$\{\varphi \ne 0\} \subset B_r \subset \{\eta = 0\} \subset \mathcal{N} = \{\eta < 1\} \subset B_R,$$

the last two inclusions being strict. We define the competitor

$$v_n = u_n + \varphi + (1 - \eta)(u_\infty - u_n),$$

and we set for simplicity $v_\infty := u_\infty + \varphi$. Now, since $\varphi = 0$ on $B_R \setminus \mathcal{N}$, we have that:

- $v_n = v_\infty$ on the set $\{\eta = 0\}$;
- (3.17) is equivalent to $\displaystyle\int_{\mathcal{N}} |\nabla u_\infty|^2 \, dx \le \int_{\mathcal{N}} |\nabla(u_\infty + \varphi)|^2 \, dx$.

Now, using the strong H^1 convergence of u_n in $\mathcal{N}$, then the optimality of u_n and again the strong H^1 convergence from claim (i), we get

$$\int_{\mathcal{N}} |\nabla u_\infty|^2 \, dx = \lim_{n \to \infty} \int_{\mathcal{N}} |\nabla u_n|^2 \, dx \le \liminf_{n \to \infty} \int_{\mathcal{N}} |\nabla v_n|^2 \, dx = \int_{\mathcal{N}} |\nabla v_\infty|^2 \, dx,$$

which concludes the proof. $\square$

Proposition 3.15 (Lipschitz Continuity of u) *Suppose that the function $u \in H^1(B_2)$ is such that:*

(a) u is non-negative in B_2 and $u(0) = 0$;
(b) u is harmonic in $\Omega_u = \{u > 0\}$;
(c) u satisfies the minimality condition

$$\mathcal{F}_\Lambda(u) \le \mathcal{F}_\Lambda(v) \quad \text{for every} \quad v \in H^1(B_2)$$

$$\text{such that} \quad u - v \in H_0^1(B_2) \quad \text{and} \quad u \le v \text{ in } B_2.$$

Then, there is a constant C_Λ, depending only on Λ and d, such that

$$\|u\|_{L^\infty(B_{1/8})} \le C_\Lambda.$$

Proof Let $u_k \in H^1(B_2)$ be a sequence of functions satisfying the hypotheses (a), (b) and (c) above. Suppose, that $u_k(0) = 0$ and set $m_k := \|u_k\|_{L^\infty(B_{1/8})}$, for $k \ge 1$.

Fig. 3.1 The two sets $\mathcal{W}_k$
and $\Omega_k = \{u_k > 0\}$

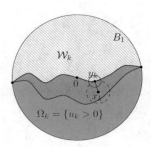

For every $k \geq 1$, we define the set (see Fig. 3.1)

$$\mathcal{W}_k := \left\{x \in B_1 \;:\; \mathrm{dist}(x, \{u_k = 0\}) \leq \frac{1}{3}(1 - |x|)\right\}.$$

Notice that, the set $\mathcal{W}_k$ and the function u_k have the following properties:

- $B_{1/8} \subset \mathcal{W}_k$ (this is due to the fact that $u_k(0) = 0$);
- u_k is continuous on $\overline{B}_1$;
- as a consequence of the previous points, we have that the maximum of u_k on the (closed) set $\mathcal{W}_k$ is achieved at a point $x_k \in \mathcal{W}_k \cap B_1$ and we have

$$M_k := u_k(x_k) = \max_{x \in \mathcal{W}_k} u_k(x) \geq m_k.$$

Let $\Omega_k := \{u_k > 0\}$ and $y_k \in \partial\Omega_k$ be the projection of x_k on the (closed) set $\partial\Omega_k \cap \overline{B}_1$. By definition $x_k \in \mathcal{W}_k$, we have that

$$r_k := |x_k - y_k| = \mathrm{dist}(x_k, \partial\Omega_k) \leq \frac{1}{3}(1 - |x_k|).$$

Thus, we get

$$|y_k| \leq |x_k| + |x_k - y_k| \leq |x_k| + \frac{1}{3}(1 - |x_k|) = 1 - \frac{2}{3}(1 - |x_k|).$$

This implies that $|y_k| < 1$ and

$$|y_k| \leq 1 - 2r_k \qquad \text{and} \qquad \frac{2}{3}r_k \leq \frac{1}{3}(1 - |y_k|).$$

Notice that the last inequality implies that $B_{r_k/2}(y_k) \subset \mathcal{W}_k$. Indeed, for every $x \in B_{r_k/2}(y_k)$, we have

$$\mathrm{dist}(x, \partial\Omega_k) \leq |x - y_k| \leq \frac{1}{2}r_k,$$

while

$$\frac{1}{3}(1 - |x|) \geq \frac{1}{3}(1 - |y_k|) - \frac{1}{3}|x - y_k| \geq \frac{2}{3}r_k - \frac{1}{6}r_k = \frac{1}{2}r_k.$$

In particular, we obtain that

$$\sup_{B_{r_k/2}(y_k)} u_k \leq M_k. \qquad (3.18)$$

On the other hand, the function u_k is harmonic in $B_{r_k}(x_k)$ so, by the Harnack inequality, we get that

$$u_k(z_k) \geq \frac{u_k(x_k)}{C_d} = \frac{M_k}{C_d} \qquad \text{where} \qquad z_k := \frac{1}{8}x_k + \frac{7}{8}y_k, \qquad (3.19)$$

and $C_d > 1$ is a dimensional constant. Now, (3.19) and (3.18) give

$$\frac{M_k}{C_d} \leq u_k(z_k) \leq \Lambda_k \leq M_k \qquad \text{where} \qquad \Lambda_k := \sup_{B_{\rho_k}(y_k)} u_k \qquad \text{and} \qquad \rho_k = \frac{r_k}{4}.$$

Consider the function

$$v_k(x) = \frac{u_k(y_k + \rho_k x)}{u_k(z_k)}.$$

and the point $\zeta_k = \dfrac{z_k - y_k}{\rho_k}$. We have that:

(1) v_k satisfies the minimality condition

$$\mathcal{F}_0(v_k) \leq \mathcal{F}_0(\phi) + \frac{\Lambda|B_2|}{u_k^2(\zeta_k)},$$

for every $\phi \in H^1(B_2)$ such that $v_k - \phi \in H_0^1(B_2)$ and $v_k \leq \phi$ in B_2;

(2) $v_k(0) = 0$ and the point $\zeta_k \in B_1$ is such that

$$|\zeta_k| = \frac{1}{2}, \qquad v_k(\zeta_k) = 1 \qquad \text{and} \qquad \sup_{B_2} v_k \leq C_d v_k(\zeta_k) = C_d;$$

(3) v_k is harmonic in $B_{1/2}(\zeta_k)$ and in Ω_{v_k};

(4) v_k is non-negative and subharmonic in B_2.

Now, by Proposition 3.13, we have that the sequence v_k is uniformly bounded in $H^1(B_1)$ and converges uniformly to a function v_∞ in B_1. Thus, we have

$$v_\infty(0) = 0 \qquad \text{and} \qquad v_\infty(\zeta_\infty) = 1 \qquad \text{and} \qquad \zeta_\infty = \lim_{k \to \infty} \zeta_k. \qquad (3.20)$$

We will next prove that v_∞ is harmonic in B_1. Let $k \in \mathbb{N}$ be fixed and let $\phi_k : B_1 \to \mathbb{R}$ be a non-negative function such that $\phi_k = v_k$ on ∂B_1. Then, since v_k is harmonic in Ω_{v_k}, we have

$$\int_{B_1} |\nabla v_k|^2 \, dx \le \int_{B_1} |\nabla(v_k \wedge \phi_k)|^2 \, dx.$$

On the other hand, the optimality condition (1) implies that

$$\int_{B_1} |\nabla v_k|^2 \, dx \le \int_{B_1} |\nabla(v_k \vee \phi_k)|^2 \, dx + \frac{\Lambda |B_2|}{u_k^2(z_k)}.$$

Putting together these two estimates, we get

$$\int_{B_1} |\nabla v_k|^2 \le \int_{B_1} |\nabla \phi_k|^2 \, dx + \varepsilon_k \qquad \text{where} \qquad \varepsilon_k := \frac{\Lambda |B_2|}{u_k^2(z_k)}.$$

Now, since $\varepsilon_k \to 0$, by Proposition 3.14, we get that v_∞ is harmonic in B_1. This is a contradiction with (3.20). □

Chapter 4
Non-degeneracy of the Local Minimizers

In this section we prove the non-degeneracy of the solutions to the one-phase problem (2.1). Our main result is the following:

Proposition 4.1 (Non-degeneracy of the Solutions: Alt-Caffarelli [3]) *Suppose that D is a bounded open set in $\mathbb{R}^d$ and $u \in H^1(D)$ is non-negative and minimizes the functional $\mathcal{F}_\Lambda$ in D, for some $\Lambda > 0$. Then, there is a constant $\kappa > 0$, depending on Λ and d, such that the following claim holds:*

$$\text{If} \quad B_r(x_0) \subset D \quad \text{and} \quad x_0 \in \overline{\Omega}_u, \quad \text{then} \quad \|u\|_{L^\infty(B_r(x_0))} \geq \kappa r.$$

The non-degeneracy holds in particular for functions satisfying the following optimality condition:

$$\mathcal{F}_\Lambda(u, \Omega) \leq \mathcal{F}_\Lambda(v, \Omega) \quad \text{for every} \quad v \in H^1(\Omega) \quad \text{such that} \quad v \leq u. \tag{4.1}$$

For the sake of completeness, we notice that this optimality condition can also be expressed in a different way, at least when it comes to functions u, which are harmonic on their positivity set Ω_u. In fact, the following result is analogous to Lemma 3.3. Moreover, as in Lemma 3.3 (see Remark 3.4), we can suppose that all the test functions v in (4.1), (4.3) and (4.2) are non-negative.

Lemma 4.2 *Let D be a bounded open set in $\mathbb{R}^d$ and $u \in H^1(D)$ be a given non-negative function. Then the following are equivalent:*

(i) u satisfies the inwards minimality condition

$$\mathcal{F}_\Lambda(u, D) \leq \mathcal{F}_\Lambda(v, D) \quad \text{for every} \quad v \in H^1(D) \quad \text{such that}$$

$$u - v \in H_0^1(D) \quad \text{and} \quad \Omega_u \supset \Omega_v. \tag{4.2}$$

© The Author(s) 2023
B. Velichkov, *Regularity of the One-phase Free Boundaries*,
Lecture Notes of the Unione Matematica Italiana 28,
https://doi.org/10.1007/978-3-031-13238-4_4

(ii) u is harmonic in Ω_u in the following sense:

$$\int_D |\nabla u|^2\, dx \le \int_D |\nabla v|^2\, dx \quad \text{for every} \quad v \in H^1(D) \quad \text{such that}$$

$$u - v \in H^1(\mathbb{R}^d) \quad \text{and} \quad u - v = 0 \quad \text{a.e. in} \quad \mathbb{R}^d \setminus \Omega_u,$$
(4.3)

and satisfies the minimality condition (4.1).

Proof The implication *(i)* $\Rightarrow$ *(ii)* is immediate. In fact, (4.2) implies both (4.3) and (4.1). In order to prove that *(ii)* implies *(i)*, we suppose that u satisfies (4.3) and (4.1) and we consider a (non-negative) function $v \in H^1(D)$ such that $u - v \in H_0^1(D)$ and $\Omega_u \subset \Omega_v$. As in the proof of Lemma 3.3, we consider the test functions $u \wedge v$ and $u \vee v$. Since $u \vee v = 0$ on $D \setminus \Omega_u$, the harmonicity of u (4.3) implies that

$$\int_D |\nabla u|^2\, dx \le \int_D |\nabla (u \vee v)|^2\, dx.$$

On the other hand, we can use $u \wedge v$ as a test function in (4.2). Thus

$$\int_D |\nabla u|^2\, dx + \Lambda|\Omega_u| \le \int_D |\nabla (u \wedge v)|^2\, dx + \Lambda|\Omega_{u \wedge v}|.$$

Summing these inequalities and using that $\Omega_v = \Omega_{u \wedge v}$, we obtain

$$2\int_D |\nabla u|^2\, dx + \Lambda|\Omega_u| \le \int_D |\nabla (u \vee v)|^2\, dx + \int_D |\nabla (u \wedge v)|^2\, dx + \Lambda|\Omega_{u \wedge v}|$$

$$= \int_D |\nabla u|^2\, dx + \int_D |\nabla v|^2\, dx + \Lambda|\Omega_v|,$$

which concludes the proof of (4.1). $\square$

Remark 4.3 (On the Terminology: Inwards Optimality and Subsolutions; Outwards Optimality and Supersolutions) We will often call the optimality conditions (4.2) and (3.1) inwards and outwards optimality condition, respectively. This is justified by the fact that the admissible test functions in (4.2) and (3.1) have positivity sets contained in or containing Ω_u. On the other hand, we will call (4.1) and (3.3) suboptimality condition and superoptimality condition, respectively, and the functions satisfying (4.1) and (3.3) will be called subsolutions and supersolutions. The terms *inwards optimality* and *outwards optimality* come from Geometric Analysis. The term *subsolution* was introduced in Shape Optimization by Bucur [8], originally to indicate inwards optimality with respect to shape functionals. The term *supersolution* appeared in the same context in several works (see for instance [51] and the references therein) to indicate outwards optimality. In the context of the functional $\mathcal{F}_\Lambda$, it seems more appropriate to use the terms *subsolution* and *supersolution*, when the condition is on the test functions, and the terms *inwards* and *outwards*, when the condition is on their (superlevel) sets. Nevertheless, being partially justified by Lemmas 3.3 and 4.2, we will often abuse this terminology by using *subsolution* and *inwards-minimizing*, and *supersolution* and *outwards-minimizing* as synonyms.

We will give two different proofs of the non-degeneracy. Lemma 4.4 is due to Alt and Caffarelli (see [3]), while Lemma 4.5 is due to David and Toro, it requires the function to be Lipschitz continuous, but the argument is more versatile and can be easily adapted, for instance, to the case of almost-minimizers of the functional $\mathcal{F}_\Lambda$ (see [19]).

Lemma 4.4 (Non-degeneracy: Alt-Caffarelli) *Let $D \subset \mathbb{R}^d$ be a bounded open set. Suppose that $u \in H^1(D)$ satisfies the condition* (4.1) *and let $x_0 \in D$. If $x_0 \in \overline{\Omega}_u \cap D$, then for every ball $B_r(x_0) \subset D$, we have that $\|u\|_{L^\infty(B_r(x_0))} \geq \Lambda^{1/2} c_d \, r$, where $c_d > 0$ is a dimensional constant.*

Proof Without loss of generality we can suppose that $x_0 = 0$ and that $\Lambda = 1$. For $r > 0$, let ϕ_r be the solution of

$$\Delta \phi_r = 0 \quad \text{in} \quad B_{2r} \setminus B_r, \qquad \phi_r = 0 \quad \text{on} \quad \partial B_r, \qquad \phi_r = 1 \quad \text{on} \quad \partial B_{2r}.$$

Then we have $\phi_r(x) = \phi_1(x/r)$, for every $x \in \overline{B}_{2r} \setminus B_r$. We consider the function $\tilde{u} \in H^1_{loc}(\Omega)$ defined by

$$\tilde{u}(x) = \begin{cases} u(x), & \text{if} \quad x \in \Omega \setminus B_{2r}, \\ u(x) \wedge M_{2r}\phi_r, & \text{if} \quad x \in B_{2r} \setminus B_r, \\ 0, & \text{if} \quad x \in B_r, \end{cases}$$

where $M_{2r} = \|u\|_{L^\infty(B_{2r})}$. By the optimality of u in B_{2r}, we have that

$$\mathcal{F}_1(u, B_{2r}) \leq \mathcal{F}_1(\tilde{u}, B_{2r}),$$

which means that

$$\mathcal{F}_1(u, B_r) \leq \mathcal{F}_1(\tilde{u}, B_{2r}) - \mathcal{F}_1(u, B_{2r} \setminus B_r) = \mathcal{F}_1(\tilde{u}, B_{2r} \setminus B_r) - \mathcal{F}_1(u, B_{2r} \setminus B_r).$$

Since $\{u > 0\} = \{\tilde{u} > 0\}$ in $B_{2r} \setminus B_r$, we get that

$$\mathcal{F}_1(\tilde{u}, B_{2r} \setminus B_r) - \mathcal{F}_1(u, B_{2r} \setminus B_r) = \int_{B_{2r} \setminus B_r} |\nabla \tilde{u}|^2 \, dx - \int_{B_{2r} \setminus B_r} |\nabla u|^2 \, dx,$$

and so, we can estimate

$$\mathcal{F}_1(u, B_r) \leq \int_{B_{2r} \setminus B_r} \left(|\nabla \tilde{u}|^2 - |\nabla u|^2 \right) dx$$

$$\leq - \int_{B_{2r} \setminus B_r} |\nabla(u - \tilde{u})|^2 \, dx + \int_{B_{2r} \setminus B_r} 2\nabla \tilde{u} \cdot \nabla(\tilde{u} - u) \, dx.$$

Now, since

$$\int_{B_{2r} \setminus B_r} 2\nabla \tilde{u} \cdot \nabla(\tilde{u} - u) \, dx = \int_{\{\tilde{u} \neq u\} \cap B_{2r} \setminus B_r} 2\nabla \tilde{u} \cdot \nabla(\tilde{u} - u) \, dx,$$

by the definition of $\tilde{u}$, we obtain

$$\mathcal{F}_1(u, B_r) \leq \int_{\{u > M_{2r}\phi_r\} \cap B_{2r} \setminus B_r} 2M_{2r} \nabla \phi_r \cdot \nabla(M_{2r}\phi_r - u) \, dx$$

$$= 2M_{2r} \int_{\partial B_r} |\nabla \phi_r| u \, d\mathcal{H}^{d-1} \, dx = 4 \frac{M_{2r}}{2r} \|\nabla \phi_1\|_{L^\infty(\partial B_r)} \int_{\partial B_r} u \, d\mathcal{H}^{d-1} \, dx.$$

On the other hand, we have the following trace inequality

$$\int_{\partial B_r} u \, d\mathcal{H}^{d-1} \leq C_d \left(\int_{B_r} |\nabla u| \, dx + \frac{1}{r} \int_{B_r} u \, dx \right)$$

$$\leq C_d \left(\int_{B_r} |\nabla u|^2 \, dx + \left(1 + \frac{M_r}{r}\right) |\{u > 0\} \cap B_r| \right) \leq C_d \left(1 + \frac{M_r}{r}\right) \mathcal{F}_1(u, B_r).$$

Thus, if $\mathcal{F}_1(u, B_r) > 0$, then we have

$$1 \leq C_d \left(1 + \frac{M_r}{r}\right) \frac{M_{2r}}{2r},$$

which gives the claim.   □

Lemma 4.5 (Non-degeneracy: David-Toro) *Suppose that $D \subset \mathbb{R}^d$ is a bounded open set and $u : D \to \mathbb{R}$ is a non-negative Lipschitz continuous functions satisfying the optimality condition (4.1). Then, there is a constant $\kappa_0 > 0$, depending on the dimension d, the Lipschitz constant $L = \|\nabla u\|_{L^\infty(D)}$ and the constant Λ, such that:*

If $x_0 \in D$ and $r \in (0, \text{dist}(x_0, \partial D))$ are such that $\fint_{\partial B_r(x_0)} u \, d\mathcal{H}^{d-1} \leq \kappa_0 r$,

   then $u = 0$ in $B_{r/8}(x_0)$.

Proof The proof is a consequence of the following three claims:

Claim 1 *Suppose that* $\fint_{\partial B_r(x_0)} u \, d\mathcal{H}^{d-1} \leq \kappa_0 r$. *Then,*

$$u \leq \kappa_1 r \quad \text{on} \quad B_{r/2}(x_0) \quad \text{where} \quad \kappa_1 = 2^d \kappa_0.$$

Claim 2 *Suppose that* $u \leq \kappa_1 r$ *on* $B_{r/2}(x_0)$. *Then,*

$$\left| \Omega_u \cap B_{r/2}(x_0) \right| \leq \kappa_2 |B_r| \quad \text{where} \quad \kappa_2 = \frac{6L + 9\kappa_1}{\Lambda} \kappa_1.$$

Claim 3 *Suppose that* $\left| \Omega_u \cap B_{r/2}(x_0) \right| \leq \kappa_2 |B_r|$ *and* $\|u\|_{L^\infty(B_{r/2}(x_0))} \leq \kappa_1 r$. *Then, for every* $y_0 \in B_{r/8}(x_0)$, *there is* $\rho \in [r/4, r/8]$ *such that*

$$\fint_{\partial B_\rho(y_0)} u \, d\mathcal{H}^{d-1} \leq \kappa_3 \rho \quad \text{where} \quad \kappa_3 = 8^{d+1} \kappa_1 \kappa_2.$$

We first prove Claim 1. Let h be the harmonic extension of u in the ball $B_r(x_0)$. By the strong maximum principle, we have that $u \leq h$ on $B_r(x_0)$ (we notice that the optimality condition (4.1) trivially implies that the function u is subharmonic). On the other hand, the Poisson formula implies that

$$h(y) = \frac{r^2 - |y|^2}{d\omega_d r} \int_{\partial B_r(x_0)} \frac{u(\zeta)}{|y - \zeta|^d} \, d\mathcal{H}^{d-1}(\zeta) \leq 2^d \kappa_0 r,$$

which gives Claim 1.

In order to prove Claim 2, we consider the function $\phi \in C_c^\infty(B_r)$ such that

$$0 \leq \phi \leq 1 \quad \text{on} \quad B_r(x_0), \qquad \phi = 1 \quad \text{on} \quad B_{r/2}(x_0), \qquad |\nabla \phi| \leq 3r^{-1}.$$

Consider the competitor $v = (u - \kappa_1 r \phi)_+$. Then, the optimality of u in $B_r(x_0)$ implies that

$$\Lambda |\Omega_u \cap B_{r/2}(x_0)| \leq \Lambda |\Omega_u \cap B_r(x_0)| - \Lambda |\Omega_v \cap B_r(x_0)| \leq \int_{B_r(x_0)} |\nabla v|^2 \, dx - \int_{B_r(x_0)} |\nabla u|^2 \, dx$$

$$\leq \int_{B_r(x_0)} |\nabla (u - \kappa_1 r \phi)|^2 \, dx - \int_{B_r(x_0)} |\nabla u|^2 \, dx$$

$$\leq 2\kappa_1 r \int_{B_r(x_0)} |\nabla u| \, |\nabla \phi| \, dx + \kappa_1^2 r^2 \int_{B_r(x_0)} |\nabla \phi|^2 \, dx \leq \left(6\kappa_1 L + 9\kappa_1^2 \right) |B_r|,$$

which concludes the proof of Claim 2.

Let us now prove Claim 3. We first estimate

$$\int_{B_{r/2}(x_0)} u \, dx \le \|u\|_{L^\infty(B_{r/2}(x_0))} |\Omega_u \cap B_{r/2}(x_0)| \le \kappa_1 \kappa_2 |B_r| r.$$

Now, taking $y_0 \in B_{r/8}(x_0)$, we have $B_{r/4}(y_0) \setminus B_{r/8}(y_0) \subset B_{r/2}(x_0)$, so there is ρ such that $r/8 \le \rho \le r/4$ and

$$\int_{\partial B_\rho(y_0)} u \, d\mathcal{H}^{d-1} \le \frac{8}{r} \int_{r/8}^{r/4} \int_{\partial B_s(y_0)} u \, d\mathcal{H}^{d-1} \, ds \le \frac{8}{r} \int_{B_{r/2}(x_0)} u \, dx$$

$$\le 8\kappa_1\kappa_2 |B_r| \le 8^{d+1} \kappa_1 \kappa_2 \omega_d \rho^d,$$

which concludes the proof of Claim 3.

We are now in position to conclude the proof of the lemma. We first notice that

$$\kappa_3 \le 8^{d+1} \kappa_1 \kappa_2 \le 2^{7d+8} \frac{L + \kappa_0}{\Lambda} \kappa_0^2.$$

Choosing

$$\kappa_0 = \inf \left\{ 1, \frac{\Lambda}{(L+1)2^{7d+8}} \right\},$$

we get that $\kappa_3 \le \kappa_0$. In particular, if $\fint_{\partial B_r(x_0)} u \, d\mathcal{H}^{d-1} \le \kappa_0 r$, then for any $y_0 \in B_{r/8}(x_0)$ there is a sequence ρ_j, $j \le 1$, such that $\frac{r}{8} \le \rho_1 \le \frac{r}{4}$ and

$$\frac{\rho_j}{8} \le \rho_{j+1} \le \frac{\rho_j}{4} \quad \text{and} \quad \fint_{\partial B_{\rho_j}(y_0)} u \, d\mathcal{H}^{d-1} \le \kappa_0 \rho_j \qquad \text{for every} \qquad j \ge 1.$$

In particular, this implies that $u = 0$ in $B_{r/8}(x_0)$, which proves the claim. $\square$

Chapter 5
Measure and Dimension of the Free Boundary

This chapter is dedicated to the measure theoretic structure of the free boundary $\partial\Omega_u$. The results presented here are mainly a consequence of the Lipschitz continuity and the non-degeneracy of the minimizer u (Theorem 3.1 and Proposition 4.1). The chapter is organized as follows:

- **Section 5.1.** *Density estimates for the domain Ω_u.*

    This section is dedicated to the density estimate of Ω_u at the boundary $\partial\Omega_u$. The argument presented here is precisely the one from the original work of Alt and Caffarelli [3].

- **Section 5.2.** *The positivity set Ω_u has finite perimeter.*

    In this section we prove that the set Ω_u has (locally) finite perimeter in the sense of De Giorgi. We will use this result, together with the density estimate of the previous section in order to prove that the singular part of the free boundary has zero $\mathcal{H}^{d-1}$ Hausdorff measure. The proof that we give here is the local counterpart of an argument proposed by Bucur in [8] for estimating the perimeter of the optimal sets for the higher eigenvalues of the Dirichlet Laplacian.

- **Section 5.3.** *Hausdorff measure of the free boundary.*

    In this section, we prove that the $\mathcal{H}^{d-1}$ measure of $\partial\Omega_u$ is (locally) finite.[1] Our argument is very general and essentially uses the Lipschitz continuity and non-degeneracy of u and the fact that the optimality condition (4.1) implies that Ω_u has a finite inner Minkowski content in a sense that will be specified below.

---

[1] Notice that this is not the consequence of Sect. 5.2 as the finiteness of the (generalized) perimeter implies only that the $\mathcal{H}^{d-1}$ measure of the reduced boundary is finite.

© The Author(s) 2023
B. Velichkov, *Regularity of the One-phase Free Boundaries*,
Lecture Notes of the Unione Matematica Italiana 28,
https://doi.org/10.1007/978-3-031-13238-4_5

5.1   Density Estimates for the Domain Ω_u

In this section, we prove that if u minimizes $\mathcal{F}_\Lambda$ in a set $D \subset \mathbb{R}^d$, then the set $\Omega_u = \{u > 0\}$ satisfies lower and upper (Lebesgue) density estimates at the boundary $\partial\Omega_u$. The result and the proof are due to Alt and Caffarelli [3].

Lemma 5.1 (Density Estimate) *Let* $D \subset \mathbb{R}^d$ *be a bounded open set. Let* $u : D \to \mathbb{R}$ *be a non-negative function such that:*

(a) *u is Lipschitz continuous and* $L := \|\nabla u\|_{L^\infty(D)}$;
(b) *u is non-degenerate, that is, there is a constant* $\kappa_0 > 0$ *such that*

$$\fint_{\partial B_r(x_0)} u \, d\mathcal{H}^{d-1} \ge \kappa_0 r \qquad \text{for every} \quad x_0 \in D \cap \partial\Omega_u$$

and every $r \in (0, dist(x_0, \partial D))$;

(c) *u is subharmonic in D;*
(d) *there is* $\Lambda > 0$ *such that u satisfies the optimality condition* (3.3), *that is,*

$$\mathcal{F}_\Lambda(u, D) \le \mathcal{F}_\Lambda(v, D) \qquad \text{for every} \quad v \in H^1(D) \quad \text{such that} \quad v \ge u.$$

There is a constant $\delta_0 \in (0, 1)$, *depending on the dimension d, the Lipschitz constant L and the non-degeneracy constant* κ_0, *such that*

$$\delta_0|B_r| \le \left|\Omega_u \cap B_r(x_0)\right| \le (1 - \delta_0)|B_r|, \tag{5.1}$$

for every $x_0 \in D \cap \partial\Omega_u$ *and every* $r \in (0, dist(x_0, \partial D))$. *In particular,* (5.1) *holds for every local minimizer of* $\mathcal{F}_\Lambda$ *in D.*

Remark 5.2 Notice that the conditions (b) and (c) are fulfilled by any function satisfying the suboptimality condition (4.1). All the conditions (a), (b), (c) and (d) are satisfied for functions that minimize $\mathcal{F}_\Lambda$ in an open set $\mathcal{U}$ containing the compact set $\overline{D}$.

Proof of Lemma 5.1 Without loss of generality we can suppose that $x_0 = 0$.

We first prove the estimate by below in (5.1). Indeed, since $0 \in \partial\Omega_u$, the non-degeneracy condition (b) implies that $\|u\|_{L^\infty(B_{r/2})} \ge \kappa_0 \frac{r}{2}$. Thus, there is a point $y \in B_{r/2}$ such that $u(y) \ge \kappa_0 \frac{r}{2}$. Now, the Lipschitz continuity of u implies that $u > 0$ on the ball $B_\rho(y)$, where $\rho = \frac{r}{2} \min\left\{1, \frac{\kappa_0}{L}\right\}$, and so, we get the first estimate in (5.1).

For the upper bound on the density, we consider the harmonic replacement h of u in the ball B_r. Since u is subharmonic, we get that $u \le h$ in B_r. Now, the optimality condition (3.1), implies that

$$\Lambda\left|\{u = 0\} \cap B_r\right| \ge \int_{B_r} |\nabla u|^2 \, dx - \int_{B_r} |\nabla h|^2 \, dx = \int_{B_r} |\nabla (u - h)|^2 \, dx.$$

By the Poincaré inequality on the ball B_r we have that

$$\int_{B_r} |\nabla (h - u)|^2 \, dx \ge \frac{C_d}{r^2} \int_{B_r} |h - u|^2 \, dx \ge \frac{C_d}{|B_r|} \left(\frac{1}{r} \int_{B_r} (h - u) \, dx\right)^2.$$

The non-degeneracy of u now implies

$$h(0) = \fint_{\partial B_r} h \, d\mathcal{H}^{d-1} = \fint_{\partial B_r} u \, d\mathcal{H}^{d-1} \ge \kappa_0 \, r.$$

By the Harnack inequality applied to h, there is a dimensional constant $c_d > 0$ such that

$$h \ge c_d \, \kappa_0 \, r \quad \text{in the ball} \quad B_{r/2} \, ,$$

On the other hand, the Lipschitz continuity of u and the fact that $u(0) = 0$ give that

$$u \le L \varepsilon r \quad \text{in the ball} \quad B_{\varepsilon r} .$$

Choosing $\varepsilon > 0$ small enough such that $c_d \kappa_0 \ge 2 \varepsilon L$, we get

$$\int_{B_r} (h - u) \, dx \ge \int_{B_{\varepsilon r}} (h - u) \, dx \ge \frac{1}{2} c_d \, \kappa_0 \, r \, |B_{\varepsilon r}|,$$

which concludes the proof. $\qquad\square$

5.2 The Positivity set Ω_u Has Finite Perimeter

In this section we prove that the (generalized) perimeter of Ω_u is locally finite in D. In particular, this means that Ω_u has locally finite perimeter. The proof that we give here was already generalized in two different contexts: for the vectorial Bernoulli problem (see [42]) and for a shape optimization problem with drift (see [46]). In fact, our proof is inspired by the global argument of Bucur (see [8]) used in the context of a shape optimization problem in $\mathbb{R}^d$. The main result of this subsection is the following:

Proposition 5.3 (Inwards-Minimizing Sets Have Locally Finite Perimeter)
Suppose that D is a bounded open set in $\mathbb{R}^d$ and that $u \in H^1(D)$ is non-negative and satisfies the following minimality condition:

$$\mathcal{F}_\Lambda(u, D) \le \mathcal{F}_\Lambda(v, D) \quad \text{for every } v \in H^1(D)) \text{ such that } v \le u \text{ in } D$$

and $u - v \in H_0^1(D)$.

Then Ω_u has locally finite perimeter in D.

As a direct consequence, we obtain that the support Ω_u of a minimizer u of $\mathcal{F}_\Lambda$ has locally finite perimeter.

Corollary 5.4 (Minimizers have Locally Finite Perimeter) *Suppose that D is a bounded open set in $\mathbb{R}^d$ and that the non-negative function $u \in H^1(D)$ is a minimizer of $\mathcal{F}_\Lambda$ in D. Then Ω_u has locally finite perimeter in D.*

We divide the proof of Proposition 5.3 in two main steps: Lemmas 5.5 and 5.6. Lemma 5.5 is a sufficient condition for the local finiteness of the perimeter of a super-level set of a Sobolev function, while in Lemma 5.6, we will show that the subsolutions satisfy this condition. The conclusion of the proof of Proposition 5.3 is given at the end of the subsection.

Lemma 5.5 *Suppose that $D \subset \mathbb{R}^d$ is an open set and that $\phi : D \to [0, +\infty]$ is a function in $H^1(D)$ for which there exist $\bar{\varepsilon} > 0$ and $C > 0$ such that*

$$\int_{\{0 < \phi \le \varepsilon\} \cap D} |\nabla \phi|^2 \, dx + \Lambda \big|\{0 < \phi \le \varepsilon\} \cap D\big| \le C\varepsilon, \quad \text{for every } 0 < \varepsilon \le \bar{\varepsilon}.$$

$$(5.2)$$

Then, $Per(\{\phi > 0\}; D) \le C\sqrt{\Lambda}$.

Proof By the co-area formula, the Cauchy-Schwarz inequality and (5.2), we have that, for every $\varepsilon \le \bar{\varepsilon}$,

$$\int_0^\varepsilon \mathcal{H}^{d-1}\big(\{\phi = t\} \cap D\big) \, dt = \int_{\{0 < \phi \le \varepsilon\} \cap D} |\nabla \phi| \, dx$$

$$\le \big|\{0 < \phi \le \varepsilon\} \cap D\big|^{1/2} \left(\int_{\{0 < \phi \le \varepsilon\} \cap D} |\nabla \phi|^2 \, dx \right)^{1/2} \le \varepsilon \, C\sqrt{\Lambda}.$$

Taking $\varepsilon = 1/n$, we get that there is $\delta_n \in [0, 1/n]$ such that

$$\mathcal{H}^{d-1}\big(\partial^*\{\phi > \delta_n\} \cap D\big) \le n \int_0^{1/n} \mathcal{H}^{d-1}\big(\{\phi = t\} \cap D\big) \, dt \le C\sqrt{\Lambda}.$$

Passing to the limit as $n \to \infty$, we obtain

$$\mathcal{H}^{d-1}\big(\partial^*\{\phi > 0\} \cap D\big) \le C\sqrt{\Lambda},$$

which concludes the proof of the lemma. $\square$

Lemma 5.6 *Suppose that* $u \in H^1(B_{2r}(x_0))$ *is non-negative and satisfies the following minimality condition in the ball* $B_{2r}(x_0) \subset \mathbb{R}^d$:

$$\mathcal{F}_\Lambda(u) \le \mathcal{F}_\Lambda(v) \quad \textit{for every} \quad v \in H^1(B_{2r}(x_0)) \quad \textit{such that} \quad \begin{cases} v \le u & \textit{in} \quad B_{2r}(x_0), \\ u = v & \textit{on} \quad \partial B_{2r}(x_0). \end{cases}$$

Then, there exists a constant $C > 0$ *such that*

$$\int_{\{0 < u \le \varepsilon\} \cap B_r(x_0)} |\nabla u|^2\, dx + \Lambda \big|\{0 < u \le \varepsilon\} \cap B_r(x_0)\big| \le C\varepsilon$$

$$\textit{for every} \qquad 0 < \varepsilon \le 1. \tag{5.3}$$

Precisely, one can take

$$C = C_d\left(r^{-1}\|\nabla u\|_{L^2(B_{2r}(x_0))} + r^{-2}\right),$$

where C_d *is a dimensional constant.*

Proof We fix a function $\phi \in C^\infty(\mathbb{R}^d)$ such that

$$\phi = 0 \quad \text{in} \quad B_r \qquad \text{and} \qquad \phi = 1 \quad \text{in} \quad \mathbb{R}^d \setminus B_{2r}.$$

For a fixed $\varepsilon > 0$ we consider the functions

$$u_\varepsilon = (u - \varepsilon)_+ \qquad \text{and} \qquad \tilde{u}_\varepsilon = \phi u + (1 - \phi)u_\varepsilon.$$

We now calculate $|\nabla \tilde{u}_\varepsilon|^2$ in the ball B_{2r}.

$$\begin{aligned}
|\nabla \tilde{u}_\varepsilon|^2 &= \mathbb{1}_{\{0 < u \le \varepsilon\}}|\nabla(u\phi)|^2 + \mathbb{1}_{\{u > \varepsilon\}}|\nabla(u - \varepsilon(1 - \phi))|^2 \\
&\le \mathbb{1}_{\{0 < u < \varepsilon\}}\phi^2|\nabla u|^2 + \mathbb{1}_{\{u > \varepsilon\}}|\nabla u|^2 \\
&\quad + \varepsilon\, \mathbb{1}_{\{0 < u \le \varepsilon\}}\big(2|\nabla u||\nabla\phi| + \varepsilon|\nabla\phi|^2\big) + \varepsilon\, \mathbb{1}_{\{u > \varepsilon\}}\big(2|\nabla u||\nabla\phi| + \varepsilon|\nabla\phi|^2\big).
\end{aligned}$$

Now setting

$$C = 2\|\nabla u\|_{L^2(B_{2r})}\|\nabla\phi\|_{L^2(B_{2r})} + \|\nabla\phi\|^2_{L^2(B_{2r})},$$

and using the optimality of u, we get

$$
\begin{aligned}
0 &\geq \int_{B_{2r}} |\nabla u|^2 \, dx - \int_{B_{2r}} |\nabla \tilde{u}_\varepsilon|^2 \, dx + |\{u > 0\} \cap B_{2r}| - |\{u_\varepsilon > 0\} \cap B_{2r}| \\
&= \int_{B_{2r}} |\nabla u|^2 \, dx - \int_{B_{2r}} |\nabla \tilde{u}_\varepsilon|^2 \, dx + |\{0 < u \leq \varepsilon\} \cap B_r| \\
&\geq \int_{\{0 < u \leq \varepsilon\} \cap B_{2r}} (1 - \phi^2) |\nabla u|^2 \, dx + |\{0 < u \leq \varepsilon\} \cap B_r| - C\varepsilon \\
&\geq \int_{\{0 < u \leq \varepsilon\} \cap B_r} |\nabla u|^2 \, dx + |\{0 < u \leq \varepsilon\} \cap B_r| - C\varepsilon,
\end{aligned}
$$

which concludes the proof.   □

Proof of Proposition 5.3 Lemma 5.6 implies that (5.3) does hold. By Lemma 5.5, we obtain that the perimeter is locally bounded. Precisely,

$$
Per(\Omega_u; B_{r/2}(x_0)) \leq C \quad \text{for every } B_r(x_0) \subset D,
$$

where C depends on r, Λ and d.   □

5.3   Hausdorff Measure of the Free Boundary

In this section we prove that the $(d-1)$—dimensional Hausdorff measure of $\partial \Omega_u$ is locally finite in D. In particular, this means that Ω_u has locally finite perimeter and so, we recover Proposition 5.3. We will use the Lipschitz continuity and the non-degeneracy of the solution, as well as, the inner Hausdorff content estimate (5.4), which is a consequence of Lemma 5.6. This is a very general result, which may find application to different free boundary problems (see for instance [42]).

Proposition 5.7 *Let $D \subset \mathbb{R}^d$ be a bounded open set and $u : D \to \mathbb{R}$ a Lipschitz continuous function such that:*

(a) u is non-degenerate, that is, there is a constants $c > 0$ such that

$$
\sup_{B_r(x_0)} u \geq cr \quad \text{for every} \quad x_0 \in \partial \Omega_u \cap D \quad \text{and every} \quad 0 < r < dist(x_0, \partial D).
$$

(b) u satisfies the following (sub-)minimality condition:

$$
\mathcal{F}_\Lambda(u, D) \leq \mathcal{F}_\Lambda(v, D) \quad \text{for every } v \in H^1(D) \text{ such that } v \leq u \text{ in } D
$$

$$
\text{and } u - v \in H_0^1(D).
$$

Then, for every compact set $K \subset \Omega$, we have $\mathcal{H}^{d-1}(K \cap \partial \Omega_u) < \infty$.

As an immediate corollary, we obtain:

Corollary 5.8 (Hausdorff Measure of the Free Boundary) *Let D be a bounded open set in $\mathbb{R}^d$ and the non-negative function $u \in H^1(D)$ be a minimizer of $\mathcal{F}_\Lambda$ in D. Then, for every compact set $K \subset D$, we have $\mathcal{H}^{d-1}(K \cap \partial\Omega_u) < \infty$.*

The proof of Proposition 5.7 is a consequence of Lemma 5.6 and the following lemma.

Lemma 5.9 *Let $D \subset \mathbb{R}^d$ be an open set and $u : D \to \mathbb{R}$ a Lipschitz continuous function such that:*

(a) u is non-degenerate, that is, there is a constants $c > 0$ such that

$$\sup_{B_r(x_0)} u \geq cr \quad \text{for every} \quad x_0 \in \partial\Omega_u \cap D \quad \text{and every} \quad 0 < r < \text{dist}(x_0, \partial D).$$

(b) there is a constant $C > 0$ such that u satisfies the estimate

$$\left|\{0 < u \leq \varepsilon\} \cap D\right| \leq C\varepsilon \quad \text{for every} \quad \varepsilon > 0. \tag{5.4}$$

Then, for every compact set $K \subset \Omega$, we have $\mathcal{H}^{d-1}(K \cap \partial\Omega_u) < \infty$.

Proof Let us first recall that, for every $\delta > 0$ and every $A \subset \mathbb{R}^d$,

$$\mathcal{H}^{d-1}_{2\delta}(A) \leq \omega_{d-1} \inf\left\{ \sum_{j=1}^{\infty} r_j^{d-1} \; : \; \text{for every } B_{r_j}(x_j) \text{ such that } \bigcup_{j=1}^{\infty} B_{r_j}(x_j) \supset A \text{ and } r_j \leq \delta \right\}.$$

and

$$\mathcal{H}^{d-1}(A) = \lim_{\delta \to 0} \mathcal{H}^{d-1}_{\delta}(A).$$

Let $\delta > 0$ be fixed and let $\{B_\delta(x_j)\}_{j=1}^N$ be a covering of $K \cap \partial\Omega_u$ such that $x_j \in \partial\Omega_u$ for every $j = 1, \ldots, n$ and the balls $B_{\delta/5}(x_j)$ are disjoint. The non-degeneracy of u implies that, in every ball $B_{\delta/10}(x_j)$ there is a point y_j such that $u(y_j) \geq c\delta/10$. The Lipschitz continuity of u implies that $B_{c\delta/10L}(y_j) \subset \Omega_u$, where $L = \max\{1, \|\nabla u\|_{L^\infty}\}$. On the other hand, since $u(x_j) = 0$, we have that

$$u < L\left(\frac{c\delta}{10L} + \frac{c\delta}{10}\right) = (L+1)\frac{c\delta}{10} \quad \text{on} \quad B_{c\delta/10L}(y_j).$$

This implies that the balls $B_{c\delta/10L}(y_j)$, $j = 1, \ldots, N$, are disjoint and contained in the set $\{0 < u < (L+1)\frac{c\delta}{10}\}$. Now, the estimate from point (b) implies that

$$C(L+1)\frac{c\delta}{10} \geq \sum_{j=1}^{N} |B_{c\delta/10L}(y_j)| \geq N\omega_d \frac{c^d\delta^d}{L^d 10^d},$$

which implies that

$$N \, d\omega_d \delta^{d-1} \le dC \frac{10^{d-1}}{c^{d-1}} L^d (L+1).$$

Since, the right-hand side does not depend on δ, we get that

$$\mathcal{H}^{d-1}(K \cap \partial \Omega_u) \le dC \frac{10^{d-1}}{c^{d-1}} L^d (L+1).$$

□

Chapter 6
Blow-Up Sequences and Blow-Up Limits

Let D be an open set in $\mathbb{R}^d$ and $u : D \to \mathbb{R}$ be a (non-negative) local minimizer of $\mathcal{F}_\Lambda$ in D. Recall that, by Theorem 3.1, we have that u is locally Lipschitz continuous in D. Let $x_0 \in \partial \Omega_u \cap D$ be a given point on the free boundary. For every $r > 0$, we define the rescaled function

$$u_{x_0,r}(x) := \frac{1}{r} u(x_0 + rx).$$

Let $(r_n)_{n \geq 1}$ be a vanishing sequence of positive numbers. We say that the sequence of functions u_{x_0,r_n} is a blow-up sequence. We notice that u_{x_0,r_n} is not defined on the entire $\mathbb{R}^d$ (since a priori we might have that $D \neq \mathbb{R}^d$), its domain of definition being the set

$$\frac{1}{r}(-x_0 + D) := \{x \in \mathbb{R}^d \ : \ x_0 + rx \in D\}.$$

On the other hand, since r_n converges to zero, for every fixed $R > 0$, there exists $m > 0$ such that, for every $n \geq m$, u_{x_0,r_n} is defined on B_R, that is,

$$B_R \subset \frac{1}{r_n}(-x_0 + D).$$

Now since,

$$\nabla u_{x_0,r_n}(x) = \nabla u(x_0 + rx) \quad \text{for every} \quad x \in B_R,$$

we have that

$$\|\nabla u_{x_0,r_n}\|_{L^\infty(B_R)} = \|\nabla u\|_{L^\infty(B_{R/r_n}(x_0))}.$$

© The Author(s) 2023
B. Velichkov, *Regularity of the One-phase Free Boundaries*,
Lecture Notes of the Unione Matematica Italiana 28,
https://doi.org/10.1007/978-3-031-13238-4_6

Since u is locally Lipschitz continuous and $u(x_0) = 0$, we get that the sequence u_{x_0,r_n} is uniformly bounded and equicontinuous on B_R. Thus, by the Theorem of Ascoli-Arzelà, we obtain that there is a subsequence of u_{x_0,r_n} that converges uniformly in the ball B_R. Repeating this argument for every (natural number) $R > 0$ and extracting a diagonal sequence, we get that there exists a function $u_0 : \mathbb{R}^d \to \mathbb{R}$ such that, for every $R > 0$, the sequence u_{x_0,r_n} converges uniformly to u_0 in B_R,

$$\lim_{n \to \infty} \|u_{x_0,r_n} - u_0\|_{L^\infty(B_R)} = 0 \quad \text{for every} \quad R > 0. \tag{6.1}$$

Definition 6.1 (Blow-Up Limit) We will say that the function $u_0 : \mathbb{R}^d \to \mathbb{R}$ is a blow-up limit of u at x_0 if (6.1) does hold.

We notice that every blow-up limit u_0 of a local minimizer u of $\mathcal{F}_\Lambda$ is non-negative, Lipschitz continuous (in $\mathbb{R}^d$) and vanishes in zero. We also stress that there might be numerous blow-up limits, each one depending on the choice of the (sub-)sequence u_{x_0,r_n}. If this is the case, then we simply say that the blow-up limit is not unique. For instance, the function $\phi : B_1 \to \mathbb{R}$ defined in polar coordinates as (see Fig. 6.1)

$$\phi(\rho, \theta) = \rho \max\{0, \cos(\theta + \ln \rho)\}$$

has infinitely many blow-up limits in zero (but it is not a local minimizer of the functional $\mathcal{F}_\Lambda$). We will denote the family of all blow-up limits of u at x_0 by $\mathcal{BU}_u(x_0)$. The classification of all the possible blow-up limits and the uniqueness of the blow-up limit at a given point $x_0 \in \partial\Omega_u$ are both central questions in the free boundary regularity theory, which do not have a complete answer yet.

In this chapter we will decompose the free boundary into a regular and singular parts according to the structure of the space of blow-up limits at the points of $\partial\Omega_u$. The Sects. 6.1, 6.2, and 6.3 are dedicated to the proof of the following result.

Proposition 6.2 (Convergence of the Blow-Up Sequences) *Let D be an open subset of $\mathbb{R}^d$ and let $u : D \to \mathbb{R}$ be non-negative, $u \in H^1_{loc}(D)$ and a local minimizer of $\mathcal{F}_\Lambda$ in D. Let $x_0 \in \partial\Omega_u \cap D$ and let $r_n \to 0$ be a vanishing sequence of positive real numbers such that the blow-up sequence u_{x_0,r_n} converges locally*

Fig. 6.1 Example of a (Lipschitz) function with infinitely many blow-up limits in zero

uniformly to the blow-up limit $u_0 : \mathbb{R}^d \to \mathbb{R}$ in the sense of (6.1). Then, there is a subsequence such that, for every $R > 0$, we have:

(i) the sequence u_{x_0,r_n} converges to u_0 strongly in $H^1(B_R)$;
(ii) the sequence of characteristic functions $\mathbb{1}_{\Omega_n}$ converges to $\mathbb{1}_{\Omega_0}$ in $L^1(B_R)$, where

$$\Omega_n := \{u_{x_0,r_n} > 0\} \qquad and \qquad \Omega_0 := \{u_0 > 0\};$$

(iii) the sequence of sets $\overline{\Omega}_n$ converges locally Hausdorff in B_R to $\overline{\Omega}_0$;
(iv) u_0 is a non-trivial local minimizer of $\mathcal{F}_\Lambda$ in $\mathbb{R}^d$.

In particular, Sect. 6.1 is dedicated to the strong convergence of the blow-up sequences (claims *(i)* and *(ii)*) and the optimality of the blow-up limits (claim *(iv)*); the main result of this section (Lemma 6.3) is more general and will also be used in the proof of Theorem 1.9. Section 6.2 is dedicated to the local Hausdorff convergence of the free boundaries (claim *(iii)*); the results of this section apply both to Theorem 1.2 and Theorem 1.9. In Sect. 6.3, we conclude the proof of Proposition 6.2.

In Sect. 6.4, we define the regular part $Reg(\partial\Omega_u)$ and the singular part $Sing(\partial\Omega_u)$ of the free boundary. Moreover, we prove that the singular set $Sing(\partial\Omega_u)$ has zero $(d-1)$-dimensional Hausdorff measure (Proposition 6.12). We notice that this result applies to Theorems 1.2, 1.4, and 1.9, but is interesting only for Theorem 1.2, in which we do not make use of monotonicity formulas. In fact, in Sect. 10, we will obtain better estimates on the dimension of the singular set by means of the Weiss' monotonicity formula, which we will apply to both Theorem 1.4 and Theorem 1.9.

6.1 Convergence of Local Minimizers

In this section we prove the strong convergence of the blow-up sequences and the minimality of the blow-up limits at every point of the free boundary of a local minimizer. Our result (Lemma 6.3) is more general and applies also to other free boundary problems; for instance, we will use it in the proof of Theorem 1.9.

Lemma 6.3 *Let $\Lambda > 0$ be a given constant, $B_R \subset \mathbb{R}^d$ and $u_n \in H^1(B_R)$ be a sequence of non-negative functions such that:*

(a) every u_n is a local minimizer of $\mathcal{F}_\Lambda$ in B_R or, more generally, satisfies

$$\mathcal{F}_\Lambda(u_n, B_R) \le \mathcal{F}_\Lambda(u_n + \varphi, B_R) + \varepsilon_n \quad for\ every \quad \varphi \in H^1_0(B_r) \quad and\ every \quad r < R,$$

where ε_n is a vanishing sequence of positive constants.

(b) *the sequence u_n is uniformly bounded in $H^1(B_R)$, that is, for some constant $C > 0$,*

$$\|u_n\|_{H^1(B_R)}^2 = \mathcal{F}_0(u_n, B_R) + \int_{B_R} u_n^2 \, dx \leq C \qquad \text{for every} \qquad n \geq 1.$$

Then, there is a function $u_\infty \in H^1(B_R)$ such that, up to a subsequence, we have

(i) *u_n converges to u_∞ strongly in $H^1(B_r)$, for every $0 < r < R$;*
(ii) *the sequence of characteristic functions $\mathbb{1}_{\{u_n > 0\}}$ converges to $\mathbb{1}_{\{u_\infty > 0\}}$ strongly in $L^1(B_r)$ and pointwise almost-everywhere in B_r, for every $0 < r < R$;*
(iii) *u_∞ is a local minimizer of $\mathcal{F}_\Lambda$ in B_R.*

Proof The idea of the proof is very similar to the one in Lemma 3.14, but is more involved due to the presence of the measure term. Up to extracting a subsequence, we can suppose that the sequence u_n converges to a function $u_\infty \in H^1(B_R)$ weakly in $H^1(B_R)$, strongly in $L^2(B_R)$ and pointwise (Lebesgue) almost-everywhere in B_R. We set for simplicity

$$\Omega_n = \{u_n > 0\} \quad and \quad \Omega_\infty = \{u_\infty > 0\}.$$

The weak H^1-convergence implies that for every $0 < r \leq R$

$$\|\nabla u_\infty\|_{L^2(B_r)} \leq \liminf_{n \to \infty} \|\nabla u_n\|_{L^2(B_r)}, \tag{6.2}$$

with an equality, if and only if, (up to a subsequence) the convergence is strong in B_r. On the other hand, the pointwise convergence of u_n implies that for almost-every $x \in B_R$

$$x \in \Omega_\infty \Rightarrow u_\infty(x) > 0 \Rightarrow u_n(x) > 0 \text{ for large } n \Rightarrow x \in \Omega_n \text{ for large } n.$$

In particular, this implies that

$$\mathbb{1}_{\Omega_\infty} \leq \liminf_{n \to \infty} \mathbb{1}_{\Omega_n},$$

and so, by the Fatou Lemma, for every $0 < r \leq R$, we have

$$|\Omega_\infty \cap B_r| \leq \liminf_{n \to \infty} |\Omega_n \cap B_r|, \tag{6.3}$$

with an equality, if and only if, (again, up to a subsequence) $\mathbb{1}_{\Omega_n}$ converges strongly to $\mathbb{1}_{\Omega_\infty}$ in $L^1(B_r)$. Notice that, up to extracting a subsequence we may assume that the limits in the right-hand sides of (6.3) and (6.2) do exist.

In order to prove (i) and (ii), it is sufficient to prove that, for fixed $0 < r < R$, we have

$$\|\nabla u_\infty\|_{L^2(B_r)} = \liminf_{n \to \infty} \|\nabla u_n\|_{L^2(B_r)} \quad \text{and} \quad |\Omega_\infty \cap B_r| = \liminf_{n \to \infty} |\Omega_n \cap B_r|.$$
$$(6.4)$$

Let $\eta : B_R \to \mathbb{R}$ be a function such that

$$\eta \in C^\infty(B_R), \quad 0 \le \eta \le 1 \quad \text{in} \quad B_R, \quad \eta = 1 \quad \text{on} \quad \partial B_R, \quad \eta = 0 \quad \text{on} \quad B_r.$$
$$(6.5)$$

Consider the test function $\tilde{u}_n = \eta u_n + (1 - \eta)u_\infty$. Since u_n is a local minimizer for $\mathcal{F}_\Lambda$ in B_R, and since $u_n = \tilde{u}_n$ on ∂B_R, we have $\mathcal{F}_\Lambda(u_n, B_R) \le \mathcal{F}_\Lambda(\tilde{u}_n, B_R) + \varepsilon_n$, that is,

$$0 \le \int_{B_R} |\nabla \tilde{u}_n|^2 \, dx - \int_{B_R} |\nabla u_n|^2 \, dx + \Lambda |\tilde{\Omega}_n \cap B_R| - \Lambda |\Omega_n \cap B_R| + \varepsilon_n,$$

where we have set $\tilde{\Omega}_n := \{\tilde{u}_n > 0\}$. We first estimate

$$
\begin{aligned}
|\tilde{\Omega}_n \cap B_R| - |\Omega_n \cap B_R| &= |\tilde{\Omega}_n \cap \{\eta = 0\}| - |\Omega_n \cap \{\eta = 0\}| \\
&\quad + |\tilde{\Omega}_n \cap \{\eta > 0\}| - |\Omega_n \cap \{\eta > 0\}| \\
&= |\Omega_\infty \cap \{\eta = 0\}| - |\Omega_n \cap \{\eta = 0\}| \\
&\quad + |(\Omega_n \cup \Omega_\infty) \cap \{\eta > 0\}| - |\Omega_n \cap \{\eta > 0\}| \\
&\le |\Omega_\infty \cap \{\eta = 0\}| - |\Omega_n \cap \{\eta = 0\}| - |\{\eta > 0\}|.
\end{aligned}
$$

By the Fatou Lemma on the set $\{\eta = 0\} \setminus B_r$, we have that

$$|\Omega_\infty \cap \{\eta = 0\} \setminus B_r| \le \liminf_{n \to \infty} |\Omega_n \cap \{\eta = 0\} \setminus B_r|,$$

and so, we get

$$\limsup_{n \to \infty} \left(|\tilde{\Omega}_n \cap B_R| - |\Omega_n \cap B_R| \right) \le \limsup_{n \to \infty} \left(|\Omega_\infty \cap B_r| - |\Omega_n \cap B_r| \right) - |\{\eta > 0\}|. \quad (6.6)$$

We next calculate

$$
\begin{aligned}
|\nabla \tilde{u}_n|^2 - |\nabla u_n|^2 &= \left| \nabla(\eta u_n + (1 - \eta)u_\infty) \right|^2 - |\nabla u_n|^2 \\
&= \left| (u_n - u_\infty)\nabla \eta + \eta \nabla u_n + (1 - \eta)\nabla u_\infty \right|^2 - |\nabla u_n|^2.
\end{aligned}
$$

Now since $u_n \to u_\infty$ strongly in $L^2(B_R)$, we have that

$$\limsup_{n\to\infty} \int_{B_R} \left(|\nabla \tilde{u}_n|^2 - |\nabla u_n|^2 \right) dx$$

$$= \limsup_{n\to\infty} \int_{B_R} \left(|(u_n - u_\infty)\nabla\eta + \eta\nabla u_n + (1-\eta)\nabla u_\infty|^2 - |\nabla u_n|^2 \right) dx$$

$$= \limsup_{n\to\infty} \int_{B_R} \left((\eta^2 - 1)|\nabla u_n|^2 + 2\eta(1-\eta)\nabla u_n \cdot \nabla u_\infty + (1-\eta)^2|\nabla u_\infty|^2 \right) dx$$

$$= \limsup_{n\to\infty} \int_{B_R} (1-\eta^2)\left(|\nabla u_\infty|^2 - |\nabla u_n|^2 \right) dx$$

$$\leq \limsup_{n\to\infty} \int_{\{\eta=0\}} \left(|\nabla u_\infty|^2 - |\nabla u_n|^2 \right) dx + \int_{B_R\setminus\{\eta=0\}} |\nabla u_\infty|^2 \, dx.$$

By the weak H^1 convergence of u_n to u_∞ on the set $\{\eta = 0\} \setminus B_r$, we have

$$\limsup_{n\to\infty} \int_{B_R} \left(|\nabla \tilde{u}_n|^2 - |\nabla u_n|^2 \right) dx \leq \limsup_{n\to\infty} \int_{B_r} \left(|\nabla u_\infty|^2 - |\nabla u_n|^2 \right) dx + \int_{\{\eta>0\}} |\nabla u_\infty|^2 \, dx.$$

This estimate, together with (6.6) and the minimality of u_n, gives

$$\liminf_{n\to\infty} \mathcal{F}_\Lambda(u_n, B_r) = \liminf_{n\to\infty} \int_{B_r} |\nabla u_n|^2 \, dx + \Lambda|\Omega_n \cap B_r|$$

$$\leq \int_{B_r} |\nabla u_\infty|^2 \, dx + \Lambda|\Omega_\infty \cap B_r| + \int_{\{\eta>0\}} |\nabla u_\infty|^2 \, dx + \Lambda|\{\eta > 0\}|$$

$$= \mathcal{F}_\Lambda(u_\infty, B_r) + \int_{\{\eta>0\}} |\nabla u_\infty|^2 \, dx + \Lambda|\{\eta > 0\}|.$$

Since η is arbitrary, we finally obtain

$$\liminf_{n\to\infty} \mathcal{F}_\Lambda(u_n, B_r) \leq \mathcal{F}_\Lambda(u_\infty, B_r),$$

which implies (6.4) and, as a consequence, the claims (i) and (ii).

We now prove (iii). Let $0 < r < R$ and $\varphi \in H_0^1(B_r)$. We will show that

$$\mathcal{F}_\Lambda(u_\infty, B_r) \leq \mathcal{F}_\Lambda(u_\infty + \varphi, B_r). \tag{6.7}$$

In order to prove (6.7), we will use the optimality of u_n and we will pass to the limit. We notice that, for a fixed $n \geq 1$, the natural competitor is simply $u_n + \varphi$. Unfortunately, we cannot follow this strategy since we do NOT a priori know that

$$\lim_{n\to\infty} |\{u_n + \varphi > 0\}| = |\{u_\infty + \varphi > 0\}|.$$

Thus, we consider a function $\eta : B_R \to \mathbb{R}$ that satisfies (6.5) and is such that the set $\mathcal{N} := \{\eta < 1\}$ is a ball strictly contained in B_R. Precisely, we have that the following inclusions do hold:

$$\{\varphi \neq 0\} \subset B_r \subset \{\eta = 0\} \subset \mathcal{N} = \{\eta < 1\} \subset B_R,$$

the last two inclusions being strict. We define the competitor

$$v_n = u_n + \varphi + (1 - \eta)(u_\infty - u_n),$$

and we set for simplicity $v_\infty := u_\infty + \varphi$. Now, since $\varphi = 0$ on $B_R \setminus \mathcal{N}$, we have that $v_n = v_\infty$ on the set $\{\eta = 0\}$ and (6.7) is equivalent to

$$\mathcal{F}_\Lambda(u_\infty, \mathcal{N}) \leq \mathcal{F}_\Lambda(v_\infty, \mathcal{N}). \tag{6.8}$$

By the points (i) and (ii), we have that

$$\mathcal{F}_\Lambda(u_\infty, \mathcal{N}) = \lim_{n \to \infty} \mathcal{F}_\Lambda(u_n, \mathcal{N}).$$

The optimality of u_n and the strong H^1 convergence of u_n to u_∞ in $\mathcal{N}$ give

$$\lim_{n \to \infty} \mathcal{F}_\Lambda(u_n, \mathcal{N}) \leq \liminf_{n \to \infty} \mathcal{F}_\Lambda(v_n, \mathcal{N}) = \int_{\mathcal{N}} |\nabla v_\infty|^2 \, dx + \Lambda \liminf_{n \to \infty} |\{v_n > 0\} \cap \mathcal{N}|. \tag{6.9}$$

Moreover, since

$$v_n = v_\infty \quad \text{on the set} \quad \{\eta = 0\},$$

we have

$$|\{v_n > 0\} \cap \mathcal{N}| = |\{v_n > 0\} \cap \{\eta = 0\}| + |\{v_n > 0\} \cap \{0 < \eta < 1\}|$$
$$\leq |\{v_\infty > 0\} \cap \mathcal{N}| + |\{0 < \eta < 1\}|,$$

which, together with (6.9) and (6.8), gives

$$\mathcal{F}_\Lambda(u_\infty, \mathcal{N}) = \lim_{n \to \infty} \mathcal{F}_\Lambda(u_n, \mathcal{N}) \leq \mathcal{F}_\Lambda(v_\infty, \mathcal{N}) + |\{0 < \eta < 1\}|.$$

Now, since the set $\{0 < \eta < 1\}$ is arbitratry, we get (6.8) and so, the claim (iii). □

6.2 Convergence of the Free Boundary

This section is dedicated to the proof of Proposition 6.2 (iii). In particular, we define the notion of local Hausdorff convergence (see Definition 6.4 below) and we prove several results, which are general and can be used in the context of different free boundary problems.

Definition 6.4 (Local Hausdorff Convergence) Suppose that X_n is a sequence of closed sets in $\mathbb{R}^d$ and Ω is an open subset of $\mathbb{R}^d$. We say that X_n converges locally Hausdorff in Ω to (the closed set) X, if for every compact set $\mathcal{K} \subset \Omega$ and every open set $\mathcal{U}$, such that $\mathcal{K} \subset \mathcal{U} \subset \Omega$, we have

$$\lim_{n \to \infty} \operatorname{dist}_{\mathcal{K},\mathcal{U}}(X_n, X) = 0,$$

where, for any pair of closed subset X, Y of Ω, we define

$$\operatorname{dist}_{\mathcal{K},\mathcal{U}}(X, Y) := \max \left\{ \max_{x \in X \cap \mathcal{K}} \operatorname{dist}(x, Y \cap \mathcal{U}), \max_{y \in Y \cap \mathcal{K}} \operatorname{dist}(y, X \cap \mathcal{U}) \right\}.$$

Lemma 6.5 (Hausdorff Convergence of the Supports) *Let B_R be the ball of radius R in $\mathbb{R}^d$. Let $u_n : B_{2R} \to \mathbb{R}$ be a sequence of continuous non-negative functions such that:*

(a) u_n converges uniformly in B_{2R} to the continuous non-negative function u_0 : $B_{2R} \to \mathbb{R}$;

(b) u_n is uniformly non-degenerate, that is, there is a strictly increasing function

$$\underline{\omega} : [0, +\infty) \to [0, +\infty),$$

such that $\underline{\omega}(0) = 0$ and

$$\|u_n\|_{L^\infty(B_r(x_0))} \geq \underline{\omega}(r) \quad \text{for every} \quad x_0 \in \overline{\Omega}_{u_n} \cap B_{3R/2}, \quad r \in (0, R/2) \quad \text{and} \quad n \in \mathbb{N}.$$

Then the sequence of closed sets $\overline{\Omega}_{u_n}$ converges locally Hausdorff in B_R to $\overline{\Omega}_{u_0}$.

Proof We first prove the non-degeneracy of u_0. Suppose that $x \in \overline{\Omega}_{u_0} \cap \overline{B}_R$ and $r \leq R/2$. Then, there is $y \in B_{r/2}(x)$ such that $u_0(y) > 0$ and so, for n large enough we have that $u_n(y) > 0$. By the non-degeneracy of u_n, there is a point $z_n \in B_{r/2}(y)$ such that $u_n(z_n) \geq \underline{\omega}(r/2)$. Up to a subsequence z_n converges to some $z \in \overline{B}_{r/2}(y)$. By the uniform convergence of u_n we have

$$u_0(z) = \lim_{n \to \infty} u_n(z_n) \geq \underline{\omega}(r/2),$$

which proves that

$$\|u_0\|_{L^\infty(B_r(x))} \geq \underline{\omega}(r/2) \quad \text{for every} \quad x \in \overline{\Omega}_{u_0} \cap \overline{B}_R \quad \text{and every} \quad r \leq R/2.$$

We can now prove the local Hausdorff convergence of $\overline{\Omega}_{u_n}$ to $\overline{\Omega}_{u_0}$. Let $\mathcal{K} \subset B_R$ be a given compact set and $\mathcal{U} \subset B_R$ be an open set containing $\mathcal{K}$. Let $\delta > 0$ be the distance from $\mathcal{K}$ to the boundary of $\mathcal{U}$. We reason by contradiction. Indeed, suppose that there is $\varepsilon > 0$ such that $\mathrm{dist}_{\mathcal{K},\mathcal{U}}\big(\overline{\Omega}_{u_n}, \overline{\Omega}_{u_0}\big) > \varepsilon$. Then, up to extracting a subsequence, we can assume that one of the following does hold:

(1) There is a sequence $(x_n)_n$ such that

$$x_n \in \overline{\Omega}_{u_n} \cap \mathcal{K} \quad \text{and} \quad \mathrm{dist}\,(x_n, \overline{\Omega}_{u_0} \cap \mathcal{U}) \geq \varepsilon.$$

(2) There is a sequence $(x_n)_n$ such that

$$x_n \in \overline{\Omega}_{u_0} \cap \mathcal{K} \quad \text{and} \quad \mathrm{dist}\,(x_n, \overline{\Omega}_{u_n} \cap \mathcal{U}) \geq \varepsilon.$$

Moreover, we can assume that $0 < \varepsilon \leq \delta$.

Suppose that (1) holds. Since $x_n \in \overline{\Omega}_{u_n}$ we have that there is $y_n \in B_{\varepsilon/2}(x_n) \subset \mathcal{U}$ such that $u_n(y_n) > \underline{\omega}(\varepsilon/2)$. On the other hand, (1) implies that $u_0(y_n) = 0$, in contradiction with the uniform convergence of u_n to u_0.

Suppose that (2) holds. By the non-degeneracy of u_0 we have that there is $y_n \in B_{\varepsilon/2}(x_n) \subset \mathcal{U}$ such that $u_0(y_n) \geq \underline{\omega}(\varepsilon/4)$. On the other hand $u_n(y_n) = 0$, in contradiction with the uniform convergence of u_n to u_0. $\qquad\square$

Lemma 6.6 (Hausdorff Convergence of the Zero Level Sets) *Let B_R be the ball of radius R in $\mathbb{R}^d$. Let $u_n : B_{2R} \to \mathbb{R}$ be a sequence of continuous non-negative functions such that:*

(a) u_n converges uniformly in B_{2R} to the continuous non-negative function $u_0 : B_{2R} \to \mathbb{R}$;

(b) $u_n(0) = 0$ and u_n satisfies the following uniform growth condition:

$$u_n(x) \geq \underline{\omega}\big(\mathrm{dist}\,(x, \{u_n = 0\} \cap \overline{B}_{2R})\big) \quad \text{for every} \quad x \cap \overline{B}_R \quad \text{and every} \quad n \in \mathbb{N},$$

where $\underline{\omega} : [0, +\infty) \to [0, +\infty)$ is a strictly increasing function such that $\underline{\omega}(0) = 0$.

Then the sequence of closed sets $\{u_n = 0\}$ converges locally Hausdorff in B_R to $\{u_0 = 0\}$.

Proof Let $\mathcal{K} \subset B_R$ be a compact set and let $\mathcal{U} \subset B_R$ be an open set containing $\mathcal{K}$. Let $\delta > 0$ be the distance from $\mathcal{K}$ to the boundary $\partial \mathcal{U}$. We reason by contradiction and we suppose that there is $\varepsilon \in (0, \delta)$ such that

$$\text{dist}_{\mathcal{K}, \mathcal{U}}\left(\{u_n = 0\}, \{u_0 = 0\}\right) \geq \varepsilon.$$

Then, up to a subsequence, we have one of the following possibilities:

(1) There is a sequence $(x_n)_n$ such that

$$x_n \in \{u_n = 0\} \cap \mathcal{K} \quad \text{and} \quad \text{dist}(x_n, \{u_0 = 0\} \cap \mathcal{U}) \geq \varepsilon.$$

(2) There is a sequence $(x_n)_n$ such that

$$x_n \in \{u_0 = 0\} \cap \mathcal{K} \quad \text{and} \quad \text{dist}(x_n, \{u_n = 0\} \cap \mathcal{U}) \geq \varepsilon.$$

Suppose first that (1) holds. Up to extracting a subsequence, we can suppose that x_n converges to $x_0 \in \mathcal{K}$. By the uniform convergence of u_n and the continuity of u_0, we have

$$u_n(x_0) \leq u_n(x_n) + |u_0(x_n) - u_n(x_n)| + |u_0(x_0) - u_0(x_n)| + |u_n(x_0) - u_0(x_0)| \to 0.$$

Passing to the limit as $n \to \infty$, we get that $u_0(x_0) = 0$, which is a contradiction since

$$\text{dist}\,(x_0, \{u_0 = 0\} \cap \mathcal{U}) \geq \lim_{n \to \infty} \text{dist}\,(x_n, \{u_0 = 0\} \cap \mathcal{U}) \geq \varepsilon.$$

Suppose now that (2) holds. Now, let y_n be the point in $B_{2R} \cap \{u_n = 0\}$ that realizes the distance from x_n to this set. There are two possibilities:

- $y_n \in B_{2R} \setminus \mathcal{U}$. In this case, we have $|x_n - y_n| \geq \delta$.
- $y_n \in \mathcal{U}$. Then, we have $\text{dist}(x_n, \{u_n = 0\} \cap \mathcal{U}) = |x_n - y_n| \geq \varepsilon$.

In both cases, we have that $|x_n - y_n| \geq \varepsilon$. By the uniform growth condition (b), we have

$$u_n(x_n) \geq \underline{\omega}(|x_n - y_n|) \geq \underline{\omega}(\varepsilon),$$

which is a contradiction with the uniform convergence of u_n to u_0. $\square$

Lemma 6.7 (Hausdorff Convergence of the Free Boundaries) *Let B_R be the ball of radius R in $\mathbb{R}^d$. Let $u_n : B_{2R} \to \mathbb{R}$ be a sequence of continuous non-negative functions and $u_0 : B_{2R} \to \mathbb{R}$ be a continuous non-negative function such that:*

(a) the sequence $\overline{\Omega}_{u_n}$ converges locally Hausdorff in B_R to $\overline{\Omega}_{u_0}$;
(b) the sequence $\{u_n = 0\}$ converges locally Hausdorff in B_R to $\{u_0 = 0\}$.

Then, $\partial \Omega_{u_n}$ converges locally Hausdorff in B_R to $\partial \Omega_{u_0}$.

Proof Let $\mathcal{K} \subset B_R$ be a fixed compact set and $\mathcal{U} \subset B_R$ be a given open set. Let $\delta > 0$ be the distance between $\mathcal{K}$ and $\partial \mathcal{U}$. Let $\varepsilon \in (0, \delta)$ be fixed.

Let $x_0 \in \partial \Omega_{u_0} \cap \mathcal{K}$. By the Hausdorff convergence of $\overline{\Omega}_{u_n}$ and $\{u_n = 0\}$, we get that, for n large enough, there are points

$$y_n \in \overline{\Omega}_{u_n} \cap \mathcal{U} \qquad \text{and} \qquad z_n \in \{u_n = 0\} \cap \mathcal{U},$$

such that

$$|x_0 - y_n| < \varepsilon \qquad \text{and} \qquad |x_0 - z_n| < \varepsilon.$$

Since u_n is continuous, there is a point w_n on the segment $[y_n, z_n]$ such that $w_n \in \partial \Omega_{w_n}$. Moreover, by construction $w_n \in B_\varepsilon(x_0) \subset \mathcal{U}$. Since x_0 is arbitrary, we get that

$$\max_{x \in \partial \Omega_{u_0} \cap \mathcal{K}} \operatorname{dist}(x, \partial \Omega_{u_n} \cap \mathcal{U}) < \varepsilon.$$

Conversely, let $x_n \in \partial \Omega_{u_n} \cap \mathcal{K}$ be fixed. Using again the Hausdorff convergence of $\overline{\Omega}_{u_n}$ and $\{u_n = 0\}$, we get that, for n large enough, there are points

$$y_0 \in \overline{\Omega}_{u_0} \cap \mathcal{U} \qquad \text{and} \qquad z_0 \in \{u_0 = 0\} \cap \mathcal{U},$$

such that

$$|x_n - y_0| < \varepsilon \qquad \text{and} \qquad |x_n - z_0| < \varepsilon.$$

Now, by the continuity of u_0, we get that there is a point w_0 on the segment $[y_0, z_0]$ such that $w_0 \in \partial \Omega_{w_0} \cap B_\varepsilon(x_n)$. Since x_n is arbitrary, we get

$$\max_{x \in \partial \Omega_{u_n} \cap \mathcal{K}} \operatorname{dist}(x, \partial \Omega_{u_0} \cap \mathcal{U}) < \varepsilon,$$

which concludes the proof. $\qquad\qquad\qquad\qquad\qquad\qquad\qquad\qquad\qquad\qquad\square$

6.3 Proof of Proposition 6.2

By the local Lipschitz continuity of u, we have that for any fixed $R > 0$, the sequence $u_n = u_{x_0, r_n}$ is uniformly bounded in $H^1(B_R)$. Thus, applying Lemma 6.3, we get at once the claims (i), (ii) and (iv) of Proposition 6.2. We notice that the fact that the blow-up limit is non-trivial ($u_0 \equiv 0$) follows by the non-degeneracy of u, which assures that for every $n \geq N$ and every $R > 0$, there is a point $x_n \in \overline{B}_R$ such that $u_{x_0, r_n}(x_n) \geq \kappa$, where κ is a constant that depends only on Λ and the dimension d. The Hausdorff convergence of the free boundary (Proposition 6.2 (iii)) follows by Lemma 6.5; Lemma 6.6 and finally, by Lemma 6.7. Notice that the non-degeneracy condition of Lemma 6.5 follows by Proposition 4.1, while the uniform growth condition of Lemma 6.6 is a consequence of the following lemma (Lemma 6.8).

Lemma 6.8 Let $u : B_{2R} \to \mathbb{R}$ be a continuous non-negative function such that:

(1) $u(0) = 0$;
(2) u satisfies the following non-degeneracy condition:

$$\|u\|_{L^\infty(B_r(x))} \geq \kappa r \quad for\ every \quad x \in \overline{\Omega}_u \cap B_R \quad and\ every \quad r \in (0, R),$$

 for some given constant $\kappa > 0$;
(3) u is harmonic in $\Omega_u \cap B_{2R}$.

Then, u satisfies the following growth condition:

$$u(x) \geq \frac{\kappa}{2^{d+1}} \operatorname{dist}(x, \{u = 0\} \cap B_{2R}) \quad for\ every \quad x \in B_R .$$

Proof Suppose that $x_0 \in \Omega_u \cap B_R$ and let $y_0 \in \partial \Omega_u \cap B_{2R}$ be such that

$$r := |x_0 - y_0| = \operatorname{dist}(x_0, \{u = 0\} \cap B_{2R}).$$

Then, the non-degeneracy of u implies that there is a point $z_0 \in B_{r/2}(x_0)$ at which

$$u(z_0) \geq \kappa \frac{r}{2}.$$

Now, since u is harmonic in Ω_u, we get

$$\int_{B_{r/2}(z_0)} u(x)\, dx \geq |B_{r/2}| u(z_0) \geq \frac{\kappa\, \omega_d\, r^{d+1}}{2^{d+1}}.$$

Since u is non-negative and harmonic in $B_r(x_0)$, we have that

$$u(x_0) = \frac{1}{|B_r|} \int_{B_r(x_0)} u(x)\, dx \geq \frac{1}{|B_r|} \int_{B_{r/2}(z_0)} u(x)\, dx \geq \frac{\kappa\, \omega_d\, r^{d+1}}{\omega_d\, r^d\, 2^{d+1}} = \frac{\kappa}{2^{d+1}} r,$$

which concludes the proof. $\qquad\square$

As an immediate corollary, we obtain the following well-known result (see for instance [3]), which we give here for the sake of completeness.

Corollary 6.9 *Suppose that u is a (non-negative) minimizer of $\mathcal{F}_\Lambda$ in the ball $B_{2R} \subset \mathbb{R}^d$ such that $u(0) = 0$. Then, there are constants C_1 and C_2, depending only on Λ and d, such that the following inequality does hold:*

$$C_1 \operatorname{dist}(x, \{u = 0\} \cap B_{2R}) \leq u(x) \leq C_2 \operatorname{dist}(x, \{u = 0\} \cap B_{2R}) \quad \textit{for every } x \in B_R.$$

Proof The first inequality follows by Lemma 6.8, while the second one is due to the Lipschitz continuity of u (see Theorem 3.1). $\qquad\square$

6.4 Regular and Singular Parts of the Free Boundary

In this section, we define the regular and the singular parts of the free boundary.

We notice that we will use exactly the same definition of regular and singular parts in Theorems 1.2, 1.4, 1.9 and 1.10.

Let D be a bounded open set in $\mathbb{R}^d$ and let $u : D \to \mathbb{R}$ be a non-negative continuous function (in particular, one can take u to be a minimizer of $\mathcal{F}_\Lambda$ in D). Let x_0 be a fixed point on the free boundary $\partial\Omega_u \cap D$, where $\Omega_u = \{u > 0\}$.

Definition 6.10 (Decomposition of the Free Boundary) We say that x_0 is a regular point if there exists a blow-up limit u_0 of u at x_0 (see Definition 6.1) of the form

$$u_0(x) = \sqrt{\Lambda}\, (x \cdot \nu)_+ \quad \text{for every } x \in \mathbb{R}^d,$$

for some unit vector $\nu \in \mathbb{R}^d$. We will denote the set of all regular points $x_0 \in \partial\Omega_u \cap D$ by $Reg(\partial\Omega_u)$, and we define the singular part of the free boundary as

$$Sing(\partial\Omega_u) = (\partial\Omega_u \cap D) \setminus Reg(\partial\Omega_u).$$

In Chap. 8, we will prove that $Reg(\partial\Omega_u)$ is an open subset of $\partial\Omega_u$ and is a $C^{1,\alpha}$-regular surface in $\mathbb{R}^d$. In this section, we will prove that the reduced boundary $\partial^*\Omega_u$ is actually a subset of the regular part $Reg(\partial\Omega_u)$ and (as a consequence) that the singular set is small. Precisely, we will show that

$$\mathcal{H}^{d-1}\big(Sing(\partial\Omega_u)\big) = 0.$$

We start with the following lemma.

Lemma 6.11 *Let D be a bounded open set in $\mathbb{R}^d$ and u be a minimizer of $\mathcal{F}_\Lambda$ in D. Let $x_0 \in \partial\Omega_u \cap D$ be a free boundary point, for which there exists a unit vector $\nu \in \mathbb{R}^d$ and a vanishing sequence $r_n \to 0$ such that*

$$\mathbb{1}_{\Omega_n} \to \mathbb{1}_{H_\nu} \quad in \quad B_R \quad for\ every \quad R > 0, \tag{6.10}$$

where $\Omega_n := \frac{1}{r_n}(-x_0+\Omega_u)$ and $H_\nu := \{x \in \mathbb{R}^d \;:\; x\cdot\nu > 0\}$. Then, $x_0 \in Reg(\partial\Omega_u)$.

Proof Let u_n be the blow-up sequence

$$u_n(x) := u_{x_0,r_n}(x) = \frac{1}{r_n}u(x_0 + r_n x).$$

Notice that $\Omega_n = \{u_n > 0\}$. By Proposition 6.2, we have that, up to a subsequence and for every $R > 0$, u_n converges locally uniformly in B_R and strongly in H^1 to a function u_0, which is a non-negative Lipschitz continuous global minimizer of $\mathcal{F}_\Lambda$ in $\mathbb{R}^d$. Moreover, we have that the sequence of characteristic functions $\mathbb{1}_{\Omega_n}$ converges in $L^1(B_R)$ to $\mathbb{1}_{\Omega_{u_0}}$. In particular, this implies that $\Omega_{u_0} = H_\nu$ almost everywhere. Now, the minimality of u_0 and the fact that $|\{u_0 = 0\} \cap H_\nu| = 0$ implies that u_0 is harmonic in H_ν. By the maximum principle, we get that $\Omega_{u_0} = H_\nu$. Thus, u_0 is C^∞ smooth up to the boundary ∂H_ν (where it vanishes).

We will next prove that

$$\nabla u_0 = \sqrt{\Lambda}\,\nu \quad on \quad \partial H_\nu.$$

Indeed, suppose that this is not the case. Then, there are two possibilities:

(1) there is a point $y \in \partial H_\nu$ such that $\nabla u_0 = A\nu$ for some $A > \sqrt{\Lambda}$;
(2) there is a point $y \in \partial H_\nu$ such that $\nabla u_0 = B\nu$ for some $0 < B < \sqrt{\Lambda}$.

Suppose that (1) holds. Let $h_{r,R}$ be the radial solution from Proposition 2.15, where r is large enough and $R = R(r)$ is uniquely determined by r. Recall that:

$$r < R, \qquad \lim_{r\to\infty} |R - (r+1)| = 0, \qquad \{h_{r,R} > 0\} = B_R,$$

$$h_{r,R} = 1 \quad on \quad B_r \quad and \quad |\nabla h_{r,R}| = 1 + o(1) \quad on \quad B_R \setminus B_r.$$

Moreover, the function $\sqrt{\Lambda}\, h_{r,R}$ is a local minimizer of $\mathcal{F}_{\Lambda}$ in $\mathbb{R}^d \setminus B_r$. Let $y_r \in \mathbb{R}^d$ be such that the ball $B_R(y_r)$ is contained in H_v and is tangent to ∂H_v at y. Let $r > 0$ be fixed and such that

$$|\nabla h_{r,R}| < \frac{1}{2} + \frac{A}{2\sqrt{\Lambda}}.$$

Then, there is $\varepsilon > 0$ small enough, for which the function

$$\tilde{h}(x) := \sqrt{\Lambda}\, h_{r,R}(x + \varepsilon v)$$

satisfies the following conditions:

- the support of $\tilde{h}$ is not entirely contained in H_v, that is, $\{\tilde{h} > 0\} \cap \{u_0 = 0\} \neq \emptyset$;
- $\tilde{h} > u_0$ only in a small neighborhood of y, precisely, $\{\tilde{h} > u_0\} \subset B_{1/2}(y)$.

Next, we notice that both $\tilde{h}$ and u_0 are minimizers of $\mathcal{F}_{\Lambda}$ in $B := B_{1/2}(y)$. Since, by construction $u_0 \geq h$ on ∂B, we get that

$$\mathcal{F}_{\Lambda}(\tilde{h}, B) \leq \mathcal{F}_{\Lambda}(u_0 \wedge \tilde{h}, B) \qquad \text{and} \qquad \mathcal{F}_{\Lambda}(u_0, B) \leq \mathcal{F}_{\Lambda}(u_0 \vee \tilde{h}, B). \qquad (6.11)$$

On the other hand,

$$\mathcal{F}_{\Lambda}(\tilde{h}, B) + \mathcal{F}_{\Lambda}(u_0, B) = \mathcal{F}_{\Lambda}(u_0 \wedge \tilde{h}, B) + \mathcal{F}_{\Lambda}(u_0 \vee \tilde{h}, B),$$

which means that both the inequalities in (6.11) are equalities and that both the functions $\tilde{h} \wedge u_0$ and $\tilde{h} \vee u_0$ are minimizers of $\mathcal{F}_{\Lambda}$ in B. For instance, this means that $\tilde{h} \vee u_0$ is harmonic in the set $\{\tilde{h} > 0\} \cap B$, which is impossible since by construction $\tilde{h} \vee u_0$ is not C^1 (for instance, the gradient is not continuous on the segment $[y, y_r]$). Thus, (1) cannot happen. By the same argument, also (2) cannot happen, which means that

$$|\nabla u_0| = \sqrt{\Lambda} \quad \text{on} \quad \partial H_v.$$

Now, by the unique continuation principle we have that $u_0(x) = \sqrt{\Lambda}\,(x \cdot v)$ on H_v. Indeed, the function $\tilde{u}_0$, defined as

$$\tilde{u}_0(x) = u_0(x) \quad \text{in} \quad H_v \qquad \text{and} \qquad \tilde{u}_0(x) = \sqrt{\Lambda}\,(x \cdot v) \quad \text{on} \quad \mathbb{R}^d \setminus H_v,$$

is harmonic in the entire space $\mathbb{R}^d$ and so, it should coincide everywhere with the function $x \mapsto \sqrt{\Lambda}\,(x \cdot v)$. This concludes the proof. $\qquad\qquad\square$

Proposition 6.12 (The Singular Set Is Negligible) *Let D be a bounded open set in $\mathbb{R}^d$ and $u \in H^1(D)$ be a minimizer of $\mathcal{F}_{\Lambda}$ in D. Then, $\mathcal{H}^{d-1}\big(Sing(\partial\Omega_u)\big) = 0$.*

Proof By Proposition 5.3, Ω_u has locally finite perimeter in D. Let $\partial^*\Omega_u$ be the reduced boundary of Ω_u. It is well-known (see for instance [43, Theorem 5.15]) that, for every $x_0 \in \partial^*\Omega_u$, there is a unit vector $\nu \in \mathbb{R}^d$ such that the property (6.10) does hold. Thus, by Lemma 6.11, we have that $\partial^*\Omega_u \subset Reg(\partial\Omega_u)$. On the other hand, by the Second Theorem of Federer (see [43]), we have that

$$\mathcal{H}^{d-1}\big((\partial\Omega_u \cap D) \setminus (\Omega_u^{(1)} \cup \Omega_u^{(0)} \cup \partial^*\Omega_u)\big) = 0. \tag{6.12}$$

Recall that, by Lemma 5.1, there are no points of density 1 and 0 on the free boundary, that is,

$$(\partial\Omega_u \cap D) \cap \big(\Omega_u^{(1)} \cup \Omega_u^{(0)}\big) = \emptyset.$$

Thus, by (6.12)

$$\mathcal{H}^{d-1}\big((\partial\Omega_u \cap D) \setminus \partial^*\Omega_u\big) = 0.$$

Now, by the definition of the singular part, we have

$$Sing(\partial\Omega_u) = (\partial\Omega_u \cap D) \setminus Reg(\partial\Omega_u) \subset (\partial\Omega_u \cap D) \setminus \partial^*\Omega_u,$$

which concludes the proof. $\square$

Chapter 7
Improvement of Flatness

In this chapter, we will prove that the regular part of the free boundary $Reg(\partial\Omega_u)$ (defined in Sect. 6.4) is $C^{1,\alpha}$ regular, for every $\alpha \in (0, 1/2)$. We will first show that the minimizers of $\mathcal{F}_\Lambda$ are viscosity solutions of an overdetermined boundary value problem. Precisely, we will prove the following result.

Proposition 7.1 (Local Minimizers Are Viscosity Solutions) *Let D be a bounded open set of $\mathbb{R}^d$ and let $u \in H^1(D)$ be a minimizer of $\mathcal{F}_\Lambda$ in D. Then, u is a viscosity solution of*

$$\Delta u = 0 \quad in \quad \Omega_u, \qquad |\nabla u| = \sqrt{\Lambda} \quad on \quad \partial\Omega_u \cap D, \qquad (7.1)$$

in the sense of Definition 7.6.

The rest of the section is dedicated to the De Silva improvement of flatness theorem [23]. Precisely, we will prove that the (viscosity) solutions to (7.1) have $C^{1,\alpha}$ regular free boundary. The proof follows step-by-step (sometimes with minor modifications) the original proof of De Silva [23].

Without loss of generality, we can assume that $\Lambda = 1$. This is due to the following remark, which is an immediate consequence of the definition of a viscosity solution (Definition 7.6).

Remark 7.2 The continuous non-negative function $u : B_1 \to \mathbb{R}$ is a viscosity solution to (7.1), for some $\Lambda > 0$, if and only if the function $v := \Lambda^{-1/2} u$ is a viscosity solution to

$$\Delta v = 0 \quad in \quad \Omega_v, \qquad |\nabla v| = 1 \quad on \quad \partial\Omega_v \cap D. \qquad (7.2)$$

As a consequence, it is sufficient to give the notion of flatness in the case $\Lambda = 1$.

© The Author(s) 2023
B. Velichkov, *Regularity of the One-phase Free Boundaries*,
Lecture Notes of the Unione Matematica Italiana 28,
https://doi.org/10.1007/978-3-031-13238-4_7

Definition 7.3 (Flatness) Let $u : B_1 \to \mathbb{R}$ be a given function. Let $\varepsilon > 0$ be a fixed real number and $v \in \mathbb{R}^d$ a unit vector. We say that

$$u \text{ is } \varepsilon\text{-flat, in the direction } v, \text{ in } B_1,$$

if

$$(x \cdot v - \varepsilon)_+ \le u(x) \le (x \cdot v + \varepsilon)_+ \qquad \text{for every} \qquad x \in B_1.$$

Theorem 7.4 (Improvement of Flatness for Viscosity Solutions, De Silva [23])
There are dimensional constants $C_0 > 0$, $\varepsilon_0 > 0$, $\sigma \in (0,1)$ and $r_0 > 0$ such that the following holds:
  If $u : B_1 \to \mathbb{R}$ be a continuous function such that:

(a) u is non-negative and $0 \in \partial\Omega_u$;
(b) u is a viscosity solution to

$$\begin{cases} \Delta u = 0 & in \quad \Omega_u \cap B_1, \\ |\nabla u| = 1 & on \quad \partial\Omega_u \cap B_1; \end{cases}$$

(c) there is $\varepsilon \in (0, \varepsilon_0]$ such that u is ε-flat in B_1, in the direction of the unit vector $v \in \mathbb{R}^d$.

  Then, there is a unit vector $\tilde{v} \in \partial B_1 \subset \mathbb{R}^d$ such that:

(i) $|\tilde{v} - v| \le C_0 \varepsilon$;
(ii) the function $u_{r_0} : B_1 \to \mathbb{R}$ is $\sigma\varepsilon$-flat in B_1, in the direction $\tilde{v}$, where we recall that $u_{r_0}(x) = \dfrac{1}{r_0} u(r_0 x)$.

  Precisely, for any $\varepsilon_0 > 0$, we may take

$$C_0 = C_d, \qquad \varepsilon_0 = r_0 \qquad and \qquad \sigma = C_d r_0,$$

where C_d is a dimensional constant.

From the improvement of flatness (Theorem 7.4) we will deduce the regularity of the free boundary (see Chap. 8). The section is organized as follows:

- In Sect. 7.1 we give the definition of a viscosity solution and we prove Proposition 7.1 using as competitors the radial solutions from the Propositions 2.15 and 2.16.

- In order to prove Theorem 7.4, we will reason by contradiction. This means that we will need a compactness result for a sequence of viscosity solutions $u_n : B_1 \to \mathbb{R}$ which are ε_n-flat in B_1 (ε_n being an infinitesimal sequence). This will be the aim of Sects. 7.2 and 7.3. In Sect. 7.2, we will prove the so-called Partial Harnack inequality (see Theorem 7.7), which we will use in Sect. 7.3 to obtain the compactness result (Lemma 7.14).
- Section 7.4 is dedicated to the proof of Theorem 7.4.
- Sections 8.1 and 8.2 are dedicated to the proof of Theorem 8.1, which is based on a classical argument and is well-known to be a consequence of the improvement of flatness Theorem 7.4.

7.1 The Optimality Condition on the Free Boundary

In this section, we give the definition of a viscosity solution and we prove Proposition 7.1.

Definition 7.5 Suppose that $\Omega \subset \mathbb{R}^d$ is an open set and that u is a continuous function, defined on the closure $\overline{\Omega}$. Let $x_0 \in \overline{\Omega}$. We say that the function $\phi \in C^\infty(\mathbb{R}^d)$ touches u from below (resp. from above) at x_0 in Ω if:

- $u(x_0) = \phi(x_0)$;
- there is a neighborhood $\mathcal{N}(x_0) \subset \mathbb{R}^d$ of x_0 such that $u(x) \geq \phi(x)$ (resp. $u(x) \leq \phi(x)$), for every $x \in \mathcal{N}(x_0) \cap \overline{\Omega}$.

Definition 7.6 (Viscosity Solutions) Let $D \subset \mathbb{R}^d$ be an open set, $A > 0$ and $u : D \to \mathbb{R}^+$ be a continuous function. We say that u is a viscosity solution of the problem

$$\Delta u = 0 \quad \text{in} \quad \Omega_u, \qquad |\nabla u| = A \quad \text{on} \quad \partial\Omega_u \cap D,$$

if for every $x_0 \in \overline{\Omega}_u \cap D$ and $\phi \in C^\infty(D)$, we have

- if $x_0 \in \Omega_u = \{u > 0\}$ and

  - if ϕ touches u from below at x_0 in Ω_u, then $\Delta\phi(x_0) \leq 0$;
  - if ϕ touches u from above at x_0 in Ω_u, then $\Delta\phi(x_0) \geq 0$;

- if $x_0 \in \partial\Omega_u \cap D$ and

  - if ϕ touches u from below at x_0 in Ω_u, then $|\nabla\phi(x_0)| \leq A$;
  - if ϕ^+ touches u from above at x_0 in Ω_u, then $|\nabla\phi(x_0)| \geq A$.

Proof of Proposition 7.1 Suppose that $x_0 \in \Omega_u$ and that $\phi \in C^\infty(D)$ touches u from below in x_0. Since u is harmonic (and smooth) in the (open) set Ω_u, we get that $\Delta\phi(x_0) \geq 0$. The case when ϕ touches u from above at $x_0 \in \Omega_u$ is analogous. Let now $x_0 \in \partial\Omega_u$. Suppose that ϕ touches u from below at x_0 and that $|\nabla\phi(x_0)| > 1$. We assume that $x_0 = 0$ and we set

$$|\nabla\phi(0)| = a \qquad \text{and} \qquad v = \frac{1}{a}\nabla\phi(0) \in \partial B_1,$$

we get that, for some $\rho > 0$ small enough,

$$u(x) \geq \phi_+(x) \geq \frac{1+a}{2}(x \cdot v)_+ \qquad \text{for every} \qquad x \in B_\rho.$$

Let now $r > 0$ be large enough such that the radial solution u_r from Lemma 2.15 satisfies

$$u_r = 1 \quad \text{in} \quad B_r, \qquad u_r = 0 \quad \text{in} \quad \mathbb{R}^d \setminus B_R, \qquad |\nabla u_r| \leq \frac{2+a}{3} \quad \text{in} \quad B_R \setminus B_r.$$

Let $\widetilde{u}_\varepsilon$ be the following translation of u_r

$$\widetilde{u}_\varepsilon(x) := u_r\big(x - (R - \varepsilon)v\big).$$

Choosing ε small enough we can suppose that $\tilde{u}_\varepsilon(0) > 0$ but

$$\widetilde{u}_\varepsilon(x) \leq \frac{1+a}{2}(x \cdot v)_+ \qquad \text{for every} \qquad x \in \partial B_\rho.$$

Thus,

$$\widetilde{u}_\varepsilon \vee u = u \qquad \text{and} \qquad \widetilde{u}_\varepsilon \wedge u = \widetilde{u}_\varepsilon \quad \text{on} \quad \partial B_\rho.$$

Now, since both $\widetilde{u}_\varepsilon$ and u are both minimizers in B_ρ, we get

$$\mathcal{F}_\Lambda(\widetilde{u}_\varepsilon, B_\rho) \leq \mathcal{F}_\Lambda(\widetilde{u}_\varepsilon \wedge u, B_\rho) \qquad \text{and} \qquad \mathcal{F}_\Lambda(u, B_\rho) \leq \mathcal{F}_\Lambda(\widetilde{u}_\varepsilon \vee u, B_\rho).$$

On the other hand, we have

$$\mathcal{F}_\Lambda(\widetilde{u}_\varepsilon, B_\rho) + F_\Lambda(u, B_\rho) = \mathcal{F}_\Lambda(\widetilde{u}_\varepsilon \wedge u, B_\rho) + \mathcal{F}_\Lambda(\widetilde{u}_\varepsilon \vee u, B_\rho),$$

which gives that

$$\mathcal{F}_\Lambda(\widetilde{u}_\varepsilon, B_\rho) = \mathcal{F}_\Lambda(\widetilde{u}_\varepsilon \wedge u, B_\rho) \qquad \text{and} \qquad \mathcal{F}_\Lambda(u, B_\rho) = \mathcal{F}_\Lambda(\widetilde{u}_\varepsilon \vee u, B_\rho).$$

Now, we define the function

$$\widetilde{v}_\varepsilon = \begin{cases} \widetilde{u}_\varepsilon & \text{in} \quad \mathbb{R}^d \setminus B_\rho, \\ \widetilde{u}_\varepsilon \wedge u & \text{in} \quad B_\rho, \end{cases}$$

and we set $v_r(x) = \widetilde{v}_\varepsilon\big(x + (R - \varepsilon)v\big)$. Thus, we get that $\mathcal{F}_\Lambda(v_r, \mathbb{R}^d) = \mathcal{F}_\Lambda(u_r, \mathbb{R}^d)$, but $v_r \neq u_r$, which is a contradiction with Lemma 2.15. The case when ϕ touches u from above is analogous and follows by Lemma 2.16. $\qquad\square$

7.2 Partial Harnack Inequality

In this section we prove a weak version of Theorem 7.4. We will assume that u satisfies the conditions (a), (b) and (c) of Theorem 7.4, which means, in particular, that u is ε-flat in some direction v:

$$(x \cdot v - \varepsilon)_+ \leq u(x) \leq (x \cdot v + \varepsilon)_+ \quad \text{for every} \quad x \in B_1.$$

Then, we will prove that the flatness of u is improved in some smaller ball B_r. Precisely, we will show that

$$\big(x \cdot v - (1 - c)\varepsilon\big)_+ \leq u(x) \leq \big(x \cdot v + (1 - c)\varepsilon\big)_+ \quad \text{for every} \quad x \in B_r, \qquad (7.3)$$

for some dimensional constant $c \in (0, 1)$. There are two main differences with respect to Theorem 7.4:

- The flatness might not really be improved in the sense of Theorem 7.4 and Fig. 7.1. Indeed, (7.3) only implies that the rescaled function

$$u_r : B_1 \to \mathbb{R}, \quad u_r(x) = \frac{1}{r}u(rx),$$

is $(1 - c)\frac{\varepsilon}{r}$—flat in B_1. Since the constants c and r are small, we might have

$$(1 - c)\frac{\varepsilon}{r} \geq \varepsilon,$$

which means that u_r might not be flatter than u.

Fig. 7.1 Improvement of flatness in the ball B_1. For simplicity, we set $r := r_0$

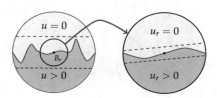

- The flatness direction does not change ($v' = v$). Notice that, without changing the direction, the improvement of flatness (in the sense of Theorem 7.4) should not hold. In fact, the function $u(x) = x_d^+$ is ε-flat in the direction v (whenever $|v - e_d| = \varepsilon$), but for any $r > 0$, $u_r(x) = u(x) = x_d^+$, thus u_r cannot be more than ε-flat in the direction v (the improvement is only possible if we are allowed to replace v by a vector, which is closer to e_d).

The main result of this section is the following.

Theorem 7.7 (Partial Boundary Harnack) *There are dimensional constants $\bar{\varepsilon} > 0$ and $c \in (0,1)$ such that for every viscosity solution u of (7.2) in $B_1 \subset \mathbb{R}^d$ such that $0 \in \overline{\Omega}_u$ we have the following property:*
If there are two real numbers $a_0 < b_0$ such that

$$|b_0 - a_0| \leq \bar{\varepsilon} \qquad and \qquad (x_d + a_0)_+ \leq u(x) \leq (x_d + b_0)_+ \qquad on \quad B_1,$$

then there are real numbers a_1 and b_1 such that $a_0 \leq a_1 < b_1 \leq b_0$,

$$|b_1 - a_1| \leq (1-c)|a_0 - b_0| \qquad and \qquad (x_d + a_1)_+ \leq u(x) \leq (x_d + b_1)_+ \qquad on \quad B_{1/20}.$$

Proof Since $0 \in \overline{\Omega}_u$ we have that $a_0 \geq -1/10$. We consider two cases:

1. Suppose that $|a_0| \leq 1/10$. Then applying Lemma 7.10 we have the claim.
2. Suppose that $a_0 \geq 1/10$. Then u is harmonic in $B_1 \cap \{x_d > -1/10\}$ (and so, in the ball $B_{1/10}$) and the claim follows by Lemma 7.9.

$\square$

We next prove the two main results: Lemmas 7.9 and 7.10. Section 7.2.1 is dedicated to the proof of Lemma 7.9, which is a consequence of the classical Harnack inequality for harmonic functions stated in Lemma 7.8. Section 7.2.2 is dedicated to the boundary version of the Harnack inequality (Lemma 7.10), which is due to De Silva [23].

7.2.1 Interior Harnack Inequality

Lemma 7.8 (Harnack Inequality) *There is a dimensional constant $C_{\mathcal{H}}$ such that for every $h : B_{2r}(x_0) \to \mathbb{R}$, a non-negative harmonic function in the ball $B_{2r}(x_0) \subset \mathbb{R}^d$, the following (Harnack) inequality does hold*

$$\max_{B_r(x_0)} h \leq C_{\mathcal{H}} \min_{B_r(x_0)} h. \tag{7.4}$$

In particular, we have

$$h(x_0) \le C_{\mathcal{H}} \min_{B_r(x_0)} h.$$

Proof The proof is an immediate consequence of the mean-value property. $\square$

Lemma 7.9 (Improvement of Flatness at Fixed Scale) *Let $C_{\mathcal{H}} > 1$ be the dimensional constant from the Harnack inequality (7.4) and let $c_{\mathcal{H}} := (2C_{\mathcal{H}})^{-1}$. Suppose that $u : B_{2r} \to \mathbb{R}$ is a harmonic function for which there are a constant $\varepsilon > 0$ and a linear function $\ell : \mathbb{R}^d \to \mathbb{R}$ such that*

$$\ell(x) \le u(x) \le \ell(x) + \varepsilon \qquad \text{for every} \qquad x \in B_{2r}.$$

Then at least one of the following does hold :

(i) $\ell(x) + c_{\mathcal{H}}\varepsilon \le u(x) \le \ell(x) + \varepsilon$ *for every* $x \in B_r$;
(ii) $\ell(x) \le u(x) \le \ell(x) + (1 - c_{\mathcal{H}})\varepsilon$ *for every* $x \in B_r$.

Proof We consider two cases.
Case 1. Suppose that $u(0) \ge \ell(0) + \varepsilon/2$. Then the function $h := u - \ell$ is harmonic and non-negative in B_{2r}. Then, by the Harnack inequality (7.4), we have

$$\frac{\varepsilon}{2} \le h(0) \le C_{\mathcal{H}} \min_{B_r} h,$$

which means that

$$u - \ell \ge \frac{\varepsilon}{2C_{\mathcal{H}}} \quad \text{in} \quad B_r,$$

and so (i) holds.
Case 2. Suppose that $u(0) \le \ell(0) + \varepsilon/2$. Then the function $h := \ell + \varepsilon - u$ is harmonic and non-negative in B_{2r}. Then, by the Harnack inequality (7.4), we have

$$\frac{\varepsilon}{2} \le h(0) \le C_{\mathcal{H}} \min_{B_r} h,$$

which means that

$$\ell + \varepsilon - u \ge \frac{\varepsilon}{2C_{\mathcal{H}}} \quad \text{in} \quad B_r,$$

and so (ii) holds. $\square$

7.2.2   Partial Harnack Inequality at the Free Boundary

Lemma 7.10 (Improving the Flatness at Fixed Scale; De Silva [23]) *There are dimensional constants $\bar{\varepsilon} > 0$ and $c \in (0, 1)$, for which the following does hold.*
   Suppose that $u : B_1 \to \mathbb{R}$ is a continuous non-negative function and a viscosity solution of (7.2) in $B_1 \subset \mathbb{R}^d$. Then, we have the following property:
   If there are real constants ε and σ, $0 < \varepsilon \le \bar{\varepsilon}$ and $|\sigma| < 1/10$, such that

$$(x_d + \sigma)_+ \le u(x) \le (x_d + \sigma + \varepsilon)_+ \qquad \textit{for every} \qquad x \in B_1,$$

then at least one of the following does hold :

(i) $(x_d + \sigma + c\varepsilon)_+ \le u(x) \le (x_d + \sigma + \varepsilon)_+$ *for every* $x \in B_{1/20}$;
(ii) $(x_d + \sigma)_+ \le u(x) \le (x_d + \sigma + (1 - c)\varepsilon)_+$ *for every* $x \in B_{1/20}$.

Proof We set

$$\bar{x} = \frac{e_d}{5} \qquad \text{and} \qquad \bar{c} = \left(20^d - \left(4/3 \right)^d \right)^{-1},$$

and consider the function $w : \mathbb{R}^d \to \mathbb{R}$, defined as (see Fig. 7.2):

$$w(x) = 1 \text{ for } x \in B_{1/20}(\bar{x}),$$

$$w(x) = 0 \text{ for } x \in \mathbb{R}^d \setminus B_{3/4}(\bar{x}),$$

$$w(x) = \bar{c} \left(|x - \bar{x}|^{-d} - \left(3/4 \right)^{-d} \right),$$

$$\text{for every } x \in B_{3/4}(\bar{x}) \setminus \overline{B}_{1/20}(\bar{x}).$$

   The set, where the function w is not constantly vanishing, is precisely the ball $B_{3/4}(\bar{x})$ (see Fig. 7.2). Moreover, on the annulus $B_{3/4}(\bar{x}) \setminus \overline{B}_{1/20}(\bar{x})$, the function w has the following properties :

(w1) $\Delta w(x) = 2 d \bar{c} |x - \bar{x}|^{-(d+2)} \ge 2 d \bar{c} \left(4/3 \right)^{d+2} > 0$.
(w2) $\partial_{x_d} w \ge C_w > 0$ on the half-space $\{x_d < 1/10\}$. Here, $C_w > 0$ is an (explicit) constant depending only on the dimension.

   Following the notation from [23] we set $p(x) = x_d + \sigma$. Similarly to what we did in the proof of Lemma 7.9, we consider two cases.
Case 1. Suppose that $u(\bar{x}) \ge p(\bar{x}) + \varepsilon/2$.
   Since the function $u - p$ is harmonic and non-negative in the ball $B_{1/10}(\bar{x})$, we can apply the Harnack inequality (7.4). Thus, setting $c_{\mathcal{H}} := \left(2 C_{\mathcal{H}} \right)^{-1}$ we get

$$u(x) - p(x) \ge c_{\mathcal{H}} \varepsilon \quad \text{in} \quad B_{1/20}(\bar{x}).$$

Fig. 7.2 The function w

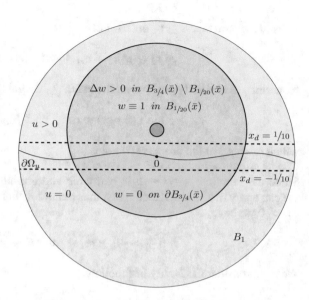

$\Delta w > 0$ in $B_{3/4}(\bar{x}) \setminus B_{1/20}(\bar{x})$

$w \equiv 1$ in $B_{1/20}(\bar{x})$

$u > 0$

$x_d = 1/10$

$\partial \Omega_u$

$x_d = -1/10$

$u = 0$ $w = 0$ on $\partial B_{3/4}(\bar{x})$

B_1

We now consider the family of functions

$$v_t(x) = p(x) + c_{\mathcal{H}}\varepsilon w(x) - c_{\mathcal{H}}\varepsilon + c_{\mathcal{H}}\varepsilon t.$$

We will prove that for every $t \in [0, 1)$, we have $u(x) \geq v_t(x)$ in B_1. We notice that, for $t < 1$ the function v_t has the following properties:

(v1) $v_t(x) < p(x) \leq u(x)$ on $B_1 \setminus B_{3/4}(\bar{x})$ (since the support of w is precisely $B_{3/4}(\bar{x})$),

(v2) $v_t(x) < u(x)$ in $\overline{B}_{1/20}(\bar{x})$ (by the choice of the constant $c_{\mathcal{H}}$),

(v3) $\Delta v_t(x) > 0$ on the blue annulus $B_{3/4}(\bar{x}) \setminus \overline{B}_{1/20}(\bar{x})$ (follows from (w1)),

(v4) $|\nabla v_t|(x) \geq \partial_{x_d} v_t(x) \geq 1 + c_{\mathcal{H}}\varepsilon C_w > 1$ on $\left(B_{3/4}(\bar{x}) \setminus \overline{B}_{1/20}(\bar{x})\right) \cap \{x_d < 1/10\}$.

Suppose (by absurd) that, for some $t \in [0, 1)$, the function $u - v_t$ has local minimum in B_1 in a point $x \in \overline{B}_1$. By (v3) and the fact that u is a viscosity solution we have that $x \notin \Omega_u \cap \left(B_{3/4}(\bar{x}) \setminus \overline{B}_{1/20}(\bar{x})\right)$. By (v4) we have that $x \notin \partial \Omega_u \cap \left(B_{3/4}(\bar{x}) \setminus \overline{B}_{1/20}(\bar{x})\right)$ and $x \notin (B_1 \setminus \overline{\Omega}_u) \cap \left(B_{3/4}(\bar{x}) \setminus \overline{B}_{1/20}(\bar{x})\right)$. Thus we get $x \notin B_{3/4}(\bar{x}) \setminus \overline{B}_{1/20}(\bar{x})$. By (v1) and (v2) we conclude that

$$\min_{x \in \overline{B}_1} \{u(x) - v_t(x)\} > 0 \qquad \text{whenever} \qquad t < 1.$$

Thus, we obtain that $u \geq v_1$ on $\overline{B}_1$, i.e.

$$u(x) \geq p(x) + c_{\mathcal{H}} \varepsilon w(x) \quad \text{on} \quad B_1.$$

Now since w is strictly positive on the ball $B_{1/20}$ we get that

$$u(x) \geq p(x) + c_d \varepsilon \quad \text{on} \quad B_{1/20},$$

which proves that the property *(i)* holds.

Case 2. Suppose that $u(\bar{x}) < p(\bar{x}) + \varepsilon/2$.

Since the function $p + \varepsilon - u$ is harmonic and non-negative in the ball $B_{1/10}(\bar{x})$, we can apply the Harnack inequality, thus obtaining that for a dimensional constant $c_{\mathcal{H}} > 0$ we have

$$p + \varepsilon - u \geq c_{\mathcal{H}} \varepsilon \quad \text{in} \quad B_{1/20}(\bar{x}).$$

We now consider the family of functions

$$v_t(x) = p(x) + \varepsilon - c_{\mathcal{H}} \varepsilon w(x) + c_{\mathcal{H}} \varepsilon - c_{\mathcal{H}} t,$$

and, reasoning as in the previous case, we get that

$$v_t(x)^+ \geq u(x) \quad \text{for every} \quad t \in [0, 1).$$

In particular, since w is strictly positive on the ball $B_{1/20}$, we get that

$$u(x) \leq \big(p(x) + (1 - c_d)\varepsilon\big)_+ \quad \text{on} \quad B_{1/20},$$

which concludes the proof. $\square$

7.3 Convergence of Flat Solutions

In this subsection we prove the compactness result that we will need in the proof of Theorem 7.4. The proof is entirely based on Theorem 7.7, from which we know that any (continuous, non-negative) viscosity solution $u : B_1 \to \mathbb{R}$ of (7.2) satisfies the following condition.

Condition 7.11 (Partial Improvement of Flatness) *There are constants $\bar{\varepsilon} > 0$ and $c \in (0, 1)$ such that the following holds. If $x_0 \in \overline{\Omega}_u$, $B_r(x_0) \subset B_1$ and $a_0 < b_0$ are such that*

$$|b_0 - a_0| \leq r\bar{\varepsilon} \quad \text{and} \quad (x_d + a_0)_+ \leq u(x) \leq (x_d + b_0)_+ \quad \text{on} \quad B_r(x_0),$$

then there are real numbers a_1 and b_1 such that $a_0 \leq a_1 < b_1 \leq b_0$,

$$|b_1 - a_1| \leq (1-c)|a_0 - b_0| \qquad and \qquad (x_d + a_1)_+ \leq u(x) \leq (x_d + b_1)_+ \qquad on \qquad B_{r/20}(x_0).$$

Remark 7.12 We notice that if $u : B_1 \to \mathbb{R}$ is a continuous non-negative function on B_1, then, for any $a < b$ and any set $E \subset B_1$, the inequality

$$(x_d + a)_+ \leq u(x) \leq (x_d + b)_+ \qquad on \quad E,$$

is equivalent to

$$x_d + a \leq u(x) \leq x_d + b \qquad on \quad E \cap \overline{\Omega}_u,$$

Thus, an equivalent way to state Condition 7.11 is the following. The non-negative function $u : B_1 \to \mathbb{R}$ satisfies Condition 7.11, if and only if, the following holds.
 If $x_0 \in \overline{\Omega}_u$, $B_r(x_0) \subset B_1$ and $a_0 < b_0$ are such that

$$|b_0 - a_0| \leq r\bar{\varepsilon} \qquad and \qquad x_d + a_0 \leq u(x) \leq x_d + b_0 \qquad on \quad B_r(x_0) \cap \overline{\Omega}_u,$$

then there are real numbers a_1 and b_1 such that $a_0 \leq a_1 < b_1 \leq b_0$,

$$|b_1 - a_1| \leq (1-c)|a_0 - b_0| \qquad and \qquad x_d + a_1 \leq u(x) \leq x_d + b_1 \qquad on \quad B_{r/20}(x_0) \cap \overline{\Omega}_u.$$

The constants $\bar{\varepsilon}$ and c are the same as in Condition 7.11.

Lemma 7.13 *Suppose that the continuous non-negative function $u : B_1 \to \mathbb{R}$ satisfies Condition 7.11 with constants c and $\bar{\varepsilon}$. Suppose that $0 \in \overline{\Omega}_u$ and that there are two real numbers $a_0 < b_0$ such that*

$$\varepsilon := |b_0 - a_0| < \frac{\bar{\varepsilon}}{2} \qquad and \qquad x_d + a_0 \leq u(x) \leq x_d + b_0 \qquad on \quad B_1 \cap \overline{\Omega}_u.$$

Then, setting

$$\tilde{u}(x) = \frac{u(x) - x_d}{\varepsilon} \qquad for\ every \qquad x \in \overline{\Omega}_u \cap B_1,$$

for every $x_0 \in B_{1/2} \cap \overline{\Omega}_u$, we have the uniform estimate

$$|\tilde{u}(x) - \tilde{u}(x_0)| \le C|x - x_0|^\gamma \qquad \text{for every} \qquad x \in \overline{\Omega}_u \cap \left(B_{1/2}(x_0) \setminus B_{\varepsilon/\bar{\varepsilon}}(x_0) \right),$$

where C is a numerical constant and γ depends only on c.

Proof Let $n \ge 0$ be such that

$$\frac{1}{2}\left(1/20\right)^{n+1} \le \frac{\varepsilon}{\bar{\varepsilon}} < \frac{1}{2}\left(1/20\right)^n.$$

Let $r_j = \frac{1}{2}\left(1/20\right)^j$. Then, we have

$$\varepsilon \le \bar{\varepsilon} r_j \qquad \text{for every} \qquad j = 0, 1, \dots, n.$$

Thus, for every $x_0 \in B_{1/2} \cap \overline{\Omega}_u$ we can apply the (partial) improvement of flatness in $B_{r_j}(x_0)$, for every $j = 0, 1, \dots, n$. Thus, we get that there are

$$a_0 \le a_1 \le \dots \le a_j \le \dots \le a_n \le b_n \le \dots \le b_j \le \dots \le b_1 \le b_0$$

such that

$$|b_j - a_j| \le (1-c)^j |a_0 - b_0| \qquad \text{and} \qquad (x_d + a_j)_+ \le u(x) \le (x_d + b_j)_+ \quad \text{on} \quad B_{r_j}(x_0),$$

which implies that

$$x_d + a_j \le u(x) \le x_d + b_j \quad \text{on} \quad B_{r_j}(x_0) \cap \overline{\Omega}_u,$$

and so,

$$\left|\left(u(x) - x_d\right) - a_j\right| \le (1-c)^j \varepsilon \quad \text{for} \quad x \in B_{r_j}(x_0) \cap \overline{\Omega}_u.$$

The triangular inequality implies that

$$|\tilde{u}(x) - \tilde{u}(x_0)| \le 2(1-c)^j \quad \text{for every} \quad x \in B_{r_j}(x_0) \cap \overline{\Omega}_u,$$

which gives the claim by choosing j such that

$$r_{j+1} < |x - x_0| \le r_j,$$

and setting γ to be such that $\left(1/20\right)^\gamma = 1 - c$.   □

Lemma 7.14 (Compactness for Flat Sequences) *Let $\bar{\varepsilon} > 0$ and $c \in (0, 1)$ be fixed constants. Suppose that $u_k : B_1 \to \mathbb{R}$ is a sequence of continuous non-negative functions such that*

(a) u_k satisfies Condition 7.11 in B_1 with constants $\bar{\varepsilon}$ and c.
(b) u_k is ε_k-flat in B_1, that is,

$$x_d - \varepsilon_k \leq u_k(x) \leq x_d + \varepsilon_k \quad in \quad B_1 \cap \overline{\Omega}_{u_k}.$$

(c) $\lim_{k \to \infty} \varepsilon_k = 0.$

Then there is a Hölder continuous function $\tilde{u} : B_{1/2} \cap \{x_d \geq 0\} \to \mathbb{R}$ and a subsequence of

$$\tilde{u}_k(x) = \frac{u_k(x) - x_d}{\varepsilon_k}, \qquad u_k : B_{1/2} \cap \overline{\Omega}_{u_k} \to \mathbb{R},$$

that we still denote by $\tilde{u}_k$ such that the following claims do hold.

 (i) For every $\delta > 0$, $\tilde{u}_k$ converges uniformly to $\tilde{u}$ on the set $B_{1/2} \cap \{x_d \geq \delta\}$.
(ii) The sequence of graphs

$$\Gamma_k = \left\{ (x, \tilde{u}_k(x)) \ : \ x \in \overline{\Omega}_{u_k} \cap B_{1/2} \right\} \subset \mathbb{R}^{d+1},$$

converges in the Hausdorff distance (in $\mathbb{R}^{d+1}$) to the graph

$$\Gamma = \left\{ (x, \tilde{u}(x)) \ : \ x \in B_{1/2} \cap \{x_d \geq 0\} \right\}.$$

Proof We first prove (i). For every $y \in B_{1/2} \cap \overline{\Omega}_{u_k}$ we have that

$$x_d - \varepsilon_k \leq u_k(x) \leq x_d + \varepsilon_k \quad \text{for every} \quad x \in B_{1/2}(y) \cap \overline{\Omega}_{u_k}.$$

Thus, by Lemma 7.13 we have that $\tilde{u}_k$ satisfies

$$|\tilde{u}_k(x) - \tilde{u}_k(y)| \leq C|x - y|^{\gamma} \quad \text{for every} \quad x \in B_{1/2}(y) \cap \overline{\Omega}_{u_k} \quad \text{such that} \quad |x - y| \geq \frac{\varepsilon_k}{\bar{\varepsilon}},$$

which, since y is arbitrary, gives

$$|\tilde{u}_k(x) - \tilde{u}_k(y)| \leq C|x - y|^{\gamma} \quad \text{for every} \quad x, y \in B_{1/2} \cap \overline{\Omega}_{u_k} \quad \text{such that} \quad |x - y| \geq \frac{\varepsilon_k}{\bar{\varepsilon}}.$$

Since, for $\varepsilon_k \leq \delta$, we have that $\{x_d \geq \delta\} \cap B_1 \subset \Omega_{u_k} \cap B_1$, we get that the sequence $\tilde{u}_k : \{x_d \geq \delta\} \cap B_{1/2} \to \mathbb{R}$ satisfies :

- $\tilde{u}_k$ is equi-bounded on $\{x_d \geq \delta\} \cap B_{1/2}$

$$-1 = \frac{(x_d - \varepsilon_k) - x_d}{\varepsilon_k} \leq \frac{u_k(x) - x_d}{\varepsilon_k} \leq \frac{(x_d + \varepsilon_k) - x_d}{\varepsilon_k} = 1;$$

- $\tilde{u}_k$ satisfies

$$\operatorname{osc}\left(\tilde{u}_k; A_{2r,r}(x_0) \cap \{x_d \geq \delta\} \cap B_{1/2}\right) \leq 2Cr^\gamma \qquad \text{for every} \qquad r \geq \frac{\varepsilon_k}{\varepsilon},$$

where, for any set $E \subset \overline{\Omega}_{u_k}$, we define:

$$\operatorname{osc}\left(\tilde{u}_k; E\right) := \sup_E \tilde{u}_k - \inf_E \tilde{u}_k,$$

and, for every $0 < r < R$, $A_{R,r}(x_0)$ is the annulus

$$A_{R,r}(x_0) = B_R(x_0) \setminus B_r(x_0).$$

Thus, by the Ascoli-Arzelà Theorem, there is a subsequence converging uniformly on the set $\{x_d \geq \delta\} \cap B_{1/2}$ to a Holder continuous function

$$\tilde{u} : \{x_d \geq \delta\} \cap B_{1/2} \to [-1, 1],$$

satisfying

$$|\tilde{u}(x) - \tilde{u}(y)| \leq C|x - y|^\gamma \qquad \text{for every} \qquad x, y \in B_{1/2} \cap \{x_d \geq \delta\}.$$

The above argument does not depend on $\delta > 0$. Thus, the function $\tilde{u}$ can be defined on the entire half-ball $\{x_d > 0\} \cap B_{1/2}$. Moreover, the constants C and γ do not depend on the choice of $\delta > 0$. This implies that we can extend $\tilde{u}$ to a Hölder continuous function

$$\tilde{u} : \{x_d \geq 0\} \cap B_{1/2} \to [-1, 1].$$

still satisfying the uniform continuity estimate

$$|\tilde{u}(x) - \tilde{u}(y)| \leq C|x - y|^\gamma \qquad \text{for every} \qquad x, y \in B_{1/2} \cap \{x_d \geq 0\}.$$

We now prove (ii). Suppose that $\tilde{x} = (x, \tilde{u}(x)) \in \Gamma$. For every $\delta > 0$, there is a point $y \in B_{1/2} \cap \{x_d > \delta/2\}$ such that $|x - y| \leq \delta$. (Notice that, if $x \in B_{1/2} \cap \{x_d > \delta/2\}$, then we can simply take $y = x$.) Then, setting $\tilde{y} = (y, \tilde{u}(y))$, we have the estimate

$$|\tilde{x} - \tilde{y}|^2 = |x - y|^2 + |\tilde{u}(x) - \tilde{u}(y)|^2 \leq \delta^2 + C^2\delta^{2\gamma}.$$

On the other hand, for every k such that $\varepsilon_k \leq \delta$, we have

$$\text{dist}(\tilde{y}, \Gamma_k) \leq |\tilde{u}(y) - \tilde{u}_k(y)| \leq \|\tilde{u} - \tilde{u}_k\|_{L^\infty(B_{1/2} \cap \{x_d > \delta/2\})}.$$

Thus, we finally obtain the estimate

$$\text{dist}(\tilde{x}, \Gamma_k) \leq \left(\delta^2 + C^2\delta^{2\gamma}\right)^{1/2} + \|\tilde{u} - \tilde{u}_k\|_{L^\infty(B_{1/2} \cap \{x_d > \delta/2\})}.$$

Let now $\tilde{x}_k = (x_k, \tilde{u}_k(x_k)) \in \Gamma_k$. Let k be such that $\varepsilon_k/\bar{\varepsilon} \leq \delta/2$. Let $y_k \in \{x_d \geq \delta\} \cap B_{1/2}$ be such that $\delta/2 \leq |x_k - y_k| \leq 2\delta$ and let $\tilde{y}_k = (y_k, \tilde{u}_k(y_k))$. Then, we have

$$|\tilde{x}_k - \tilde{y}_k|^2 = |x_k - y_k|^2 + |\tilde{u}_k(x_k) - \tilde{u}_k(y_k)|^2 \leq 4\delta^2 + 4C^2\delta^{2\gamma}.$$

Reasoning as above, we get

$$\text{dist}(\tilde{x}_k, \Gamma) \leq 2\left(\delta^2 + C^2\delta^{2\gamma}\right)^{1/2} + \|\tilde{u} - \tilde{u}_k\|_{L^\infty(B_{1/2} \cap \{x_d > \delta\})}.$$

Now, since δ is arbitrary and $\tilde{u}_k$ converges to $\tilde{u}$ uniformly on $\{x_d > \delta/2\} \cap B_{1/2}$, we get that

$$\lim_{k \to \infty} \text{dist}_{\mathcal{H}}(\Gamma_k, \Gamma) = 0.$$

$\square$

7.4 Improvement of Flatness: Proof of Theorem 7.4

In this subsection, we prove Theorem 7.4. Since, we will reason by contradiction, we will first study the limits of the sequences of (flat) viscosity solutions to (7.2) in B_1.

Lemma 7.15 (The Linearized Problem) *Suppose that $u_k : B_1 \to \mathbb{R}$ is a sequence of continuous non-negative functions such that:*

(a) for every k, u_k is a viscosity solution of

$$\Delta u_k = 0 \quad \text{in} \quad \Omega_{u_k} \cap B_1, \qquad |\nabla u_k| = 1 \quad \text{on} \quad \partial\Omega_{u_k} \cap B_1. \qquad (7.5)$$

(b) for every k, u_k is ε_k-flat in B_1 in the sense that

$$(x_d - \varepsilon_k)_+ \leq u_k(x) \leq (x_d + \varepsilon_k)_+ \qquad in \qquad B_1.$$

(c) $\lim_{k \to \infty} \varepsilon_k = 0$.

Then, up to extracting a subsequence, the sequence of functions

$$\tilde{u}_k \,:\, B_{1/2} \cap \overline{\Omega}_{u_k} \to \mathbb{R}, \qquad \tilde{u}_k(x) = \frac{u_k(x) - x_d}{\varepsilon_k},$$

converges (in the sense of Lemma 7.14 (i) and (ii)) to a Hölder continuous function

$$\tilde{u} \,:\, B_{1/2} \cap \{x_d \geq 0\} \to \mathbb{R}.$$

Moreover, we have that

(i) $\tilde{u}$ is a viscosity solution to

$$\Delta \tilde{u} = 0 \quad in \quad B_{1/2} \cap \{x_d > 0\}, \qquad \frac{\partial \tilde{u}}{\partial x_d} = 0 \quad on \quad B_{1/2} \cap \{x_d = 0\}, \qquad (7.6)$$

in the sense that

- *$\tilde{u}$ is harmonic in $B_{1/2} \cap \{x_d > 0\}$,*
- *If P is a polynomial touching $\tilde{u}$ from below (above) in a point $x_0 \in B_{1/2} \cap \{x_d = 0\}$, then $\dfrac{\partial P}{\partial x_d}(x_0) \leq 0$ $\left(\dfrac{\partial P}{\partial x_d}(x_0) \geq 0 \right)$.*

(ii) $\tilde{u} \in C^\infty \big(B_{1/2} \cap \{x_d \geq 0\} \big)$ and is a classical solution of (7.6).

Proof The existence of the limit function $\tilde{u}$ follows by Lemma 7.14.

We first prove (i). Suppose that P is a polynomial touching $\tilde{u}$ strictly from below in a point $x_0 \in B_{1/2} \cap \{x_d \geq 0\}$. Then there exists a sequence of points $x_k \in \overline{\Omega}_{u_k}$ such that P touches $\tilde{u}_k$ from below in x_k and $x_k \to x_0$ as $k \to \infty$. We consider two cases:

(1) Suppose that $x_0 \in \{x_d > 0\}$. Then there is some $\delta > 0$ such that $x_k \in \{x_d > \delta\}$, for every k large enough. Thus, $x_k \in \Omega_{u_k}$ for k large enough and so, since $\tilde{u}_k$ is harmonic in Ω_{u_k}, $\Delta P(x_k) \geq 0$. Passing to the limit as $k \to \infty$ we get $\Delta P(x_0) \geq 0$.
(2) Suppose that $x_0 \in \{x_d = 0\}$. We suppose without loss of generality that $x_0 = 0$. We consider the family of polynomials

$$P_\varepsilon(x) = P(x) + \frac{1}{\varepsilon} x_d^2 - \varepsilon x_d.$$

In a sufficiently small neighborhood of zero, we have that P_ε still touches $\tilde{u}$ (strictly) from below in 0. Moreover,

$$\Delta P_\varepsilon > 0 \quad \text{in a neighborhood of zero}, \qquad \frac{\partial P_\varepsilon}{\partial x_d}(0) = \frac{\partial P}{\partial x_d}(0) - \varepsilon.$$

Thus it is sufficient to show that for every $\varepsilon > 0$, we have $\dfrac{\partial P_\varepsilon}{\partial x_d}(0) \leq 0$. Let now $\varepsilon > 0$ be fixed. Consider the sequence of points $x_k \in \overline{\Omega}_{u_k}$ such that P_ε touches $\tilde{u}_k$ from below in x_k and $x_k \to x_0$ as $k \to \infty$. Since $\Delta P_\varepsilon(x_k) > 0$ and $\tilde{u}_k$ is harmonic in Ω_{u_k} we have that necessarily $x_k \in \partial \Omega_k$. By the definition of $\tilde{u}_k = \frac{u_k - x_d}{\varepsilon_k}$ we get that the polynomial $Q(x) = \varepsilon_k P_\varepsilon(x) + x_d$ touches u_k from below in x_k. Since u_k is a viscosity solution of (7.5), we get that

$$1 \geq |\nabla Q(x_k)|^2 \geq \left(1 + \varepsilon_k \frac{\partial P_\varepsilon}{\partial x_d}(x_k)\right)^2 = 1 + 2\varepsilon_k \frac{\partial P_\varepsilon}{\partial x_d}(x_k) + \varepsilon_k^2 \left|\frac{\partial P_\varepsilon}{\partial x_d}(x_k)\right|^2.$$

Thus, we have $\dfrac{\partial P_\varepsilon}{\partial x_d}(0) \leq 0$, which concludes the proof after letting $\varepsilon \to 1$.

We now prove (ii). We write $\mathbb{R}^d \ni x = (x', x_d)$ with $x' \in \mathbb{R}^{d-1}$ and $x_d \in \mathbb{R}$. We consider the function $w : \mathbb{R}^d \to \mathbb{R}$ defined by $\tilde{u}$ and its reflexion:

$$w(x', x_d) = \begin{cases} \tilde{u}(x', x_d), & \text{if } x_d \geq 0, \\ \tilde{u}(x', -x_d), & \text{if } x_d \leq 0. \end{cases}$$

We will prove that w is harmonic on $\mathbb{R}^d$. Suppose that P is a polynomial touching w strictly from below in a point $x_0 \in \{x_d = 0\}$. Since w is harmonic on $\{x_d \neq 0\}$ it is sufficient to prove that $\Delta P(x_0) \leq 0$. We first notice that since $w(x', x_d) = w(x', -x_d)$ then also the polynomial $P(x', -x_d)$ touches w strictly from below in x_0 and, as a consequence, so does the polynomial

$$Q(x', x_d) = \frac{P(x', x_d) + P(x', -x_d)}{2},$$

which satisfies

$$\Delta Q = \Delta P \quad \text{and} \quad \frac{\partial Q}{\partial x_d} = 0 \quad \text{on} \quad \{x_d = 0\}.$$

Consider the polynomial

$$Q_\varepsilon(x) = Q(x) + \varepsilon x \cdot e_d.$$

Then Q_ε touches w from below in a point x_ε and we have that $x_\varepsilon \to x_0$ as $\varepsilon \to 0$. We notice that necessarily $x_\varepsilon \in \{x_d \geq 0\}$. Moreover, we can rule out the case $x_\varepsilon \in \{x_d = 0\}$ since by the hypothesis on $\tilde{u}$ we have that in this case we should have

$$0 \geq \frac{\partial Q_\varepsilon}{\partial x_d}(x_\varepsilon) = \frac{\partial Q}{\partial x_d}(x_\varepsilon) + \varepsilon = \varepsilon,$$

which is impossible. Thus $x_\varepsilon \in \{x_d > 0\}$ and since $\tilde{u}$ is harmonic in $\{x_d > 0\}$ we get that

$$0 \geq \Delta Q_\varepsilon(x_\varepsilon) = \Delta Q(x_\varepsilon).$$

Passing to the limit as $\varepsilon \to 0$, we obtain that $\Delta Q(x_0) \leq 0$, which concludes the proof.   □

Lemma 7.16 (First and Second Order Estimates for Harmonic Functions) *Suppose that $h : B_R \to \mathbb{R}$ is a bounded harmonic function in B_R. Then*

$$\|\nabla h\|_{L^\infty(B_{R/2})} \leq \frac{C_d}{R} \|h\|_{L^\infty(B_R)}, \tag{7.7}$$

and

$$\left| h(x) - h(0) - x \cdot \nabla h(0) \right| \leq \frac{C_d}{R^2} |x|^2 \|h\|_{L^\infty(B_R)} \qquad \text{for every} \qquad x \in B_{R/2}, \tag{7.8}$$

where C_d is a dimensional constant.

Proof Let $x_0 \in B_{3R/4}$. Since h is harmonic in $B_{R/4}(x_0)$, we have that also $\partial_i h$ is harmonic in the same ball $B_{R/4}(x_0)$, we have

$$\partial_i h(x_0) = \frac{4^d}{\omega_d R^d} \int_{B_{R/4}(x_0)} \partial_i h(x)\, dx = \frac{4^d}{\omega_d R^d} \int_{B_{R/4}(x_0)} \operatorname{div} X\, dx,$$

where $X = (0, \ldots, h, \ldots, 0)$ is the vector with the only non-zero component being the ith one, which is precisely h. Now, the divergence theorem gives

$$\partial_i h(x_0) = \frac{4^d}{\omega_d R^d} \int_{\partial B_{R/4}(x_0)} \nu \cdot X\, d\mathcal{H}^{d-1} = \frac{4^d}{\omega_d R^d} \int_{\partial B_{R/4}(x_0)} \nu_i(x) h(x)\, d\mathcal{H}^{d-1}(x),$$

which implies that

$$\|\partial_i h\|_{L^\infty(B_{3R/4})} \leq \frac{4d}{R} \|h\|_{L^\infty(B_R)},$$

and so, we obtain (7.7). Now, by the same argument, we get that

$$\|\partial_{ij} h\|_{L^\infty(B_{R/2})} \le \frac{16d^2}{R^2} \|h\|_{L^\infty(B_R)}.$$

Let now $x \in B_{R/4}$ and set

$$f(t) = h(xt) \qquad \text{for every} \qquad t \in [0, 1].$$

Then, we have

$$h(x) - h(0) - x \cdot \nabla h(0) = f(1) - f(0) - f'(0) = \int_0^1 (1 - t) f''(t)\, dt.$$

Since

$$f'(t) = x \cdot \nabla h(xt) \qquad \text{and} \qquad f''(t) = \sum_{i,j=1}^d x_i x_j \partial_{ij} h(xt),$$

and

$$x \cdot \nabla h(xt) = \sum_{i,j=1}^d \int_0^t x_i x_j \partial_{ij} h(sx)\, ds,$$

we get precisely (7.8).   □

Proof of Theorem 7.4 We fix C_0 and r_0 to be dimensional constant which will be chosen later. In order to prove that ε_0 exists we reason by contradiction. Let $\varepsilon_n \to 0$ and let $u_n : B_1 \to \mathbb{R}$ be a sequence of continuous functions satisfying the conditions (a), (b) and (c) with ε_n. Without loss of generality, we may suppose that, for any $n \in \mathbb{N}$, u_n is ε_n flat in B_1 in the same direction e_d. Finally, we assume by contradiction that, there are no $n \in \mathbb{N}$ and a unit vector ν satisfying the following conditions:
   (i) $|\nu - e_d| \le C_0 \varepsilon$;
   (ii) the function $(u_n)_{r_0} : B_1 \to \mathbb{R}$ is $\sigma \varepsilon$-flat in B_1, in the direction ν.
By Lemma 7.14 we can suppose that the sequence

$$\tilde{u}_n(x) = \frac{u_n(x) - x_d}{\varepsilon_n} \qquad \text{for} \quad x \in B_1 \cap \overline{\Omega}_{u_n},$$

converges (in the sense of Lemma 7.14 (i) and (ii)) in $B_{1/2}$ to a smooth ($C^\infty(B_{1/2})$) function

$$\tilde{u} : B_{1/2} \cap \{x_d \ge 0\} \to [-1, 1]$$

that satisfies (7.6). We notice that

$$\tilde{u}(0) = 0 \quad \text{and} \quad \frac{\partial \tilde{u}}{\partial x_d}(0) = 0.$$

We set

$$v_i := \frac{\partial \tilde{u}}{\partial x_i}(0), \quad \text{for every} \quad i = 1, \ldots, d-1 \, ; \qquad v' := (v_1, \ldots, v_{d-1}) \in \mathbb{R}^{d-1},$$

and we re-write (7.8) as

$$v' \cdot x' - 4C_d |x|^2 \leq \tilde{u}(x) \leq v' \cdot x' + 4C_d |x|^2 \qquad \text{for every} \qquad x = (x', x_d) \in B_{1/4} \cap \{x_d \geq 0\}.$$

We now fix $r \leq 1/4$. Since the graph Γ_n of $\tilde{u}_n$ converges in the Hausdorff distance to the graph Γ of $\tilde{u}$ (see Lemma 7.14 (ii)), we have that for n large enough

$$v' \cdot x' - 8C_d r^2 \leq \tilde{u}_n(x) \leq v' \cdot x' + 8C_d r^2 \qquad \text{for every} \qquad x = (x', x_d) \in B_r \cap \overline{\Omega}_{u_n}. \tag{7.9}$$

Using the definition of $\tilde{u}_n$ we can rewrite (7.9) as

$$x_d + \varepsilon_n v' \cdot x' - \varepsilon_n 8C_d r^2 \leq u_n(x) \leq x_d + \varepsilon_n v' \cdot x' + \varepsilon_n 8C_d r^2, \tag{7.10}$$

which holds for every $x = (x', x_d) \in B_r \cap \overline{\Omega}_{u_n}$.

We define the new flatness direction v as follows:

$$v := \frac{1}{\sqrt{1 + \varepsilon_n^2 |v'|^2}} (\varepsilon_n v', 1) \in \mathbb{R}^d.$$

By definition, we have that $|v| = 1$. We next estimate the distance between v and e_d. Since both v and e_d are unit vectors, we have

$$|v - e_d|^2 = 2(1 - v \cdot e_d) = 2\left(1 - \frac{1}{\sqrt{1 + \varepsilon_n^2 |v'|^2}}\right).$$

Notice that the following elementary inequality holds:

$$1 - \frac{1}{\sqrt{1 + X}} \leq 2X \qquad \text{for every} \qquad -1/2 < X < 1/2. \tag{7.11}$$

In order to apply this inequality to $X = \varepsilon_n^2 |v'|^2$, we first check that $\varepsilon_n^2 |v'|^2 \leq 1/2$. In fact, by the definition of v' and (7.7), we have the estimate $|v'| \leq 2C_d$. Thus, for n

large enough, we have that $\varepsilon_n^2 |v'|^2 \leq 1/2$ and so, we can estimate

$$|v - e_d|^2 \leq 2|v'|^2 \varepsilon_n^2 \leq 8C_d^2 \varepsilon_n^2,$$

which proves that v satisfies (i), once we choose $C_0 = 4C_d$.

Using again the inequality (7.11) and the fact that

$$0 \leq u_n \leq \varepsilon_n + r \qquad \text{in} \qquad B_r,$$

which follows by the non-negativity and the ε_n-flatness of u_n, we get that

$$u_n - 8C_d^2 \varepsilon_n^2 (r + \varepsilon_n) \leq \frac{u_n}{\sqrt{1 + \varepsilon_n^2 |v'|^2}} \leq u_n \qquad \text{in} \qquad B_r.$$

Thus, dividing (7.10) by $\sqrt{1 + \varepsilon_n^2 |v'|^2}$, we get that

$$x \cdot v - C_d \left(\varepsilon_n^2 (r + \varepsilon_n) + \varepsilon_n r^2 \right) \leq u_n(x) \leq x \cdot v + C_d \varepsilon_n r^2,$$

for every $x = (x', x_d) \in B_r \cap \overline{\Omega}_{u_n}$, C_d being a dimensional constant. Choosing r_0 small enough and $\varepsilon_0 \leq r_0$, we get that

$$x \cdot v - \varepsilon_n r_0 \sigma \leq u_n(x) \leq x \cdot v + \varepsilon_n r_0 \sigma \qquad \text{for every} \qquad x = (x', x_d) \in B_{r_0} \cap \overline{\Omega}_{u_n},$$

and so the vector v satisfies (i) and (ii), in contradiction with the initial assumption.

$\square$

Chapter 8
Regularity of the Flat Free Boundaries

This chapter is dedicated to the regularity of the flat free boundaries. In particular, we will show how the improvement of flatness (proved in previous section) implies the $C^{1,\alpha}$ regularity of the free boundary (see Fig. 8.1). The results of this section are based on classical arguments and are well-known to the specialists in the field. The main result of the chapter is the following.

Theorem 8.1 (ε-Regularity for Viscosity Solutions)
  There are dimensional constants $\varepsilon > 0$ and $\delta > 0$ such that the following holds:
  Suppose that $u : B_1 \to \mathbb{R}$ satisfies the following conditions:

(a) u is a non-negative continuous function and a viscosity solution of (7.1) in B_1;
(b) u is ε-flat in B_1, that is,

$$(x_d - \varepsilon)_+ \le u(x) \le (x_d + \varepsilon)_+ \qquad \text{for every} \qquad x \in B_1.$$

Then, there is $\alpha > 0$ such that the free boundary $\partial\Omega_u$ is $C^{1,\alpha}$ regular in the cylinder $B'_\delta \times (-\delta, \delta)$. Precisely, there is a function $g : B'_\delta \to (-\delta, \delta)$ such that:

(i) g is $C^{1,\alpha}$ regular in the $(d-1)$-dimensional ball $B'_\delta \subset \mathbb{R}^{d-1}$;
(ii) the set $\Omega_u \cap \left(B'_\delta \times (-\delta, \delta)\right)$ is the supergraph of g, that is,

$$\Omega_u \cap \left(B'_\delta \times (-\delta, \delta)\right) = \left\{x = (x', x_d) \in B'_\delta \times (-\delta, \delta) \; : \; x_d > g(x')\right\}.$$

  Moreover, g (and so, $\partial\Omega_u$) is $C^{1,\alpha}$ regular, for any $\alpha \in (0, 1/2)$.

© The Author(s) 2023
B. Velichkov, *Regularity of the One-phase Free Boundaries*,
Lecture Notes of the Unione Matematica Italiana 28,
https://doi.org/10.1007/978-3-031-13238-4_8

Fig. 8.1 A flat free boundary

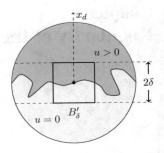

Proof The existence of a function $g : B'_\delta \subset \mathbb{R}^{d-1}$, which is $C^{1,\alpha}$ regular, for some $\alpha > 0$, for which (ii) holds, is a consequence of:

- Theorem 7.4, in which we show that the improvement of flatness (Condition 8.3) holds for viscosity solutions (with constants $\sigma = C_d \kappa$);
- Lemma 8.4, in which we show that the improvement of flatness implies the uniqueness of the blow-up limit and the decay of the blow-up sequence:

$$\|u_{r,x_0} - u_{x_0}\|_{L^\infty(B_1)} \le C_d r^\gamma \quad \text{for every} \quad r < 1/2 \quad \text{and every} \quad x_0 \in B_{1/2}, \tag{8.1}$$

  where γ is such that $\kappa^\gamma = \sigma$;
- Proposition 8.6, in which we show that if (8.1) holds, then $\partial\Omega_u$ is $C^{1,\alpha}$ regular in $B_{1/2}$, where $\alpha = \frac{\gamma}{1+\gamma}$.

In particular, we notice that by choosing κ small enough, we can take γ as close to 1 (and so, α as close to $1/2$) as we want.   □

As a consequence, we obtain the regularity of the free boundary for minimizers of $\mathcal{F}_\Lambda$.

Corollary 8.2 (Regularity of $Reg(\partial\Omega_u)$) *Let D be a bounded open set in $\mathbb{R}^d$ and let $u : D \to \mathbb{R}$ be a (non-negative) minimizer of $\mathcal{F}_\Lambda$ in D. Then, every regular point $x_0 \in Reg(\partial\Omega_u) \subset D$ has a neighborhood $\mathcal{U}$ such that $\partial\Omega_u \cap \mathcal{U}$ is a $C^{1,\alpha}$ regular manifold, for every $\alpha \in (0, 1/2)$.*

Proof Notice that, up to replacing $u(x)$ by $v(x) = \Lambda^{-1/2}u(x)$, we may assume that $\Lambda = 1$. By the definition of $Reg(\partial\Omega_u)$ (see Sect. 6.4), there is a sequence $r_n \to 0$ such that the blow-up sequence u_{r_n,x_0} converges uniformly (in B_1) to a function $u_0 : \mathbb{R}^d \to \mathbb{R}$ of the form

$$u_0(x) = (x \cdot \nu)_+$$

for some unit vector $v \in \mathbb{R}^d$. Then, by Proposition 6.2, for n large enough, we have

$$\|u_{r_n, x_0} - u_0\|_{L^\infty(B_1)} < \varepsilon,$$

$$u_{r_n, x_0} > 0 \quad \text{in} \quad \{x \cdot v > \varepsilon\} \qquad \text{and} \qquad u_{r_n, x_0} = 0 \quad \text{in} \quad \{x \cdot v < -\varepsilon\}.$$

This means that u_{r_n, x_0} is 2ε-flat in B_1, that is,

$$\left(x \cdot v - 2\varepsilon\right)_+ \le u_{r_n, x_0}(x) \le \left(x \cdot v + 2\varepsilon\right)_+ \qquad \text{for every} \qquad x \in B_1.$$

Now, taking ε small enough and applying Theorem 7.4, Proposition 7.1 and Theorem 8.1, we get the claim. $\qquad\qquad\square$

This chapter is organized as follows.

In Sect. 8.1, we prove that the improvement of flatness (Condition 8.3) implies the uniqueness of the blow-up limit and gives a (polynomial) rate of convergence of the blow-ups in $L^\infty(B_1)$.

In Sect. 8.2, we prove that the uniqueness of the blow-up limit and the polynomial rate of convergence of the blow-up sequence imply the regularity of the free boundary. We notice that the uniqueness of the blow-up limit and the rate of convergence of the blow-up sequence can be obtained also by different arguments, for instance, via an epiperimetric inequality. In fact, the result of this section can be used also in combination with Theorem 12.1, which is an alternative way to the regularity of the free boundary.

8.1 Improvement of Flatness, Uniqueness of the Blow-Up Limit and Rate of Convergence of the Blow-Up Sequence

Condition 8.3 (Improvement of Flatness) *Let* $u : B_1 \to \mathbb{R}$ *be a non-negative function. There are constants* $\kappa \in (0, 1)$, $\sigma \in (0, 1)$, $C_0 > 0$ *and* $\varepsilon_0 > 0$ *such that:*

For every $x_0 \in \partial\Omega_u \cap B_1, r \le dist(x_0, \partial B_1)$ *and* $\varepsilon \in (0, \varepsilon_0]$ *satisfying*

$$(x \cdot v - \varepsilon)_+ \le u_{r, x_0} \le (x \cdot v + \varepsilon)_+ \quad \text{in} \quad B_1,$$

there is $\tilde{v} \in \partial B_1$ *such that*

$$|\tilde{v} - v| \le C_0\varepsilon \qquad \text{and} \qquad \left(x \cdot \tilde{v} - \sigma\varepsilon\right)_+ \le u_{\kappa r, x_0} \le \left(x \cdot \tilde{v} + \sigma\varepsilon\right)_+ \quad \text{in} \quad B_1.$$

Lemma 8.4 (Uniqueness of the Blow-Up Limit) *Suppose that* $u : B_1 \to \mathbb{R}$ *is a continuous non-negative function satisfying Condition 8.3. Then, there are constant*

$\varepsilon_1 > 0$, $\gamma > 0$ and $C_1 > 0$ (depending on ε_0, κ, σ and C_0) such that if

$$(x \cdot v - \varepsilon_1)_+ \leq u \leq (x \cdot v + \varepsilon_1)_+ \quad in \quad B_1,$$

for some $v \in \partial B_1$, then for every $x_0 \in \partial \Omega_u \cap B_{1/2}$ there is a unique unit vector

$$v_{x_0} \in \partial B_1 \subset \mathbb{R}^d$$

such that

$$\|u_{r,x_0} - u_{x_0}\|_{L^\infty(B_1)} \leq C_1 r^\gamma \quad for\ every \quad r \leq 1/2,$$

where the function u_{x_0} is defined as

$$u_{x_0}(x) = (v_{x_0} \cdot x)_+ \quad for\ every \quad x \in \mathbb{R}^d.$$

Precisely, we can take γ, ε_1 and C_1 as follows:

$$\varepsilon_1 = \frac{\varepsilon_0}{4}, \qquad \kappa^\gamma = \sigma \qquad and \qquad C_1 = (2/\kappa)^\gamma \left(1 + \frac{C_0}{1-\sigma} + \frac{1}{\kappa}\right) \varepsilon_0.$$

Proof Let $\varepsilon_1 = \frac{\varepsilon_0}{4}$. Notice that if u is ε_1-flat in B_1, then

$$u_{1/2,x_0} \text{ is } \varepsilon_0\text{-flat in } B_1,$$

for every $x_0 \in \partial \Omega_u \cap B_{1/2}$.

Let $x_0 \in \partial \Omega_u \cap B_{1/2}$ be fixed,

$$r_n = \frac{\kappa^n}{2} \qquad and \qquad u_n := u_{r_n,x_0}.$$

By the improvement of flatness, there is a sequence of unit vectors $v_n \in \partial B_1$ such that

$$\left(x \cdot v_n - \varepsilon_0 \sigma^n\right)_+ \leq u_n \leq \left(x \cdot v_n + \varepsilon_0 \sigma^n\right)_+ \quad in \quad B_1,$$

and

$$|v_n - v_{n+1}| \leq C_0 \varepsilon_0 \sigma^n \quad for\ every \quad n \in \mathbb{N}.$$

In particular, for every $1 \leq n < m$, we have

$$|v_n - v_m| \leq \sum_{k=n}^{m-1} |v_k - v_{k+1}| \leq \sum_{k=n}^{m-1} C_0 \varepsilon_0 \sigma^k \leq \varepsilon C_0 \sum_{k=n}^{\infty} \sigma^k = \frac{C_0 \varepsilon}{1-\sigma} \sigma^n.$$

This implies that there is a vector $v_\infty \in \partial B_1$ such that

$$v_\infty = \lim_{n \to \infty} v_n \quad \text{and} \quad |v_n - v_\infty| \leq \sum_{k=n}^{\infty} |v_k - v_{k+1}| \leq \frac{C_0 \varepsilon_0}{1 - \sigma} \sigma^n.$$

Thus,

$$\left| x \cdot v_\infty - \left(x \cdot v_n \pm \varepsilon_0 \sigma^n \right) \right| \leq \left(1 + \frac{C_0}{1 - \sigma} \right) \varepsilon_0 \sigma^n \quad \text{for every} \quad x \in B_1,$$

which implies that

$$\left| (x \cdot v_\infty)_+ - u_n(x) \right| \leq \left(1 + \frac{C_0}{1 - \sigma} \right) \varepsilon_0 \sigma^n \quad \text{for every} \quad x \in B_1.$$

Now, we set

$$u_0(x) = (x \cdot v_\infty)_+.$$

Let $r \leq 1/2$ be arbitrary and let $n \in \mathbb{N}$ be such that

$$r_{n+1} = \frac{1}{2} \kappa^{n+1} < r \leq \frac{1}{2} \kappa^n = r_n.$$

Then, there is $\rho \in (\kappa, 1]$ such that $r = \rho r_n$. Since u_{r_n, x_0} satisfies

$$\left(x \cdot v_n - \varepsilon_0 \sigma^n \right)_+ \leq u_{r_n, x_0}(x) \leq \left(x \cdot v_n + \varepsilon_0 \sigma^n \right)_+ \quad \text{in} \quad B_1,$$

we get that $u_{r, x_0} = u_{\rho r_n, x_0}$ satisfies

$$\left(x \cdot v_n - \frac{\varepsilon_0}{\rho} \sigma^n \right)_+ \leq u_{r, x_0} \leq \left(x \cdot v_n + \frac{\varepsilon_0}{\rho} \sigma^n \right)_+ \quad \text{in} \quad B_1,$$

which implies that

$$\| u_{r_n, x_0} - u_{r, x_0} \|_{L^\infty(B_1)} \leq \frac{\varepsilon_0}{\rho} \sigma^n \leq \frac{\varepsilon_0}{\kappa} \sigma^n,$$

and finally gives that

$$\| u_{r, x_0} - u_0 \|_{L^\infty(B_1)} \leq \left(1 + \frac{C_0}{1 - \sigma} + \frac{1}{\kappa} \right) \varepsilon_0 \sigma^n.$$

Since $\kappa^\gamma = \sigma$, we get that

$$\sigma^n \leq \kappa^{n\gamma} \leq \frac{1}{\kappa^\gamma}\left(\kappa^{n+1}\right)^\gamma \leq \frac{1}{\kappa^\gamma}(2r)^\gamma = (2/\kappa)^\gamma r^\gamma,$$

from which, we deduce

$$\|u_{r,x_0} - u_0\|_{L^\infty(B_1)} \leq (2/\kappa)^\gamma \left(1 + \frac{C_0}{1-\sigma} + \frac{1}{\kappa}\right) \varepsilon_0 r^\gamma,$$

which concludes the proof.   □

8.2 Regularity of the One-Phase Free Boundaries

Condition 8.5 (Uniqueness of the Blow-Up Limit and Rate of Convergence of the Blow-Up Sequence) *The function* $u : B_1 \to \mathbb{R}$ *satisfies this condition if it is non-negative and if there are constants* $C_1 > 0$ *and* $\gamma > 0$ *such that, for every* $x_0 \in \partial\Omega_u \cap B_{1/2}$ *there is a unique function* $u_{x_0} : B_1 \to \mathbb{R}$ *such that:*

(i) there is $v_{x_0} \in \partial B_1$ *such that* $u_{x_0}(x) = (v_{x_0} \cdot x)_+$ *for every* $x \in B_1$;
(ii) $\|u_{r,x_0} - u_{x_0}\|_{L^\infty(B_1)} \leq C_1 r^\gamma$ *for every* $r \leq 1/2$.

Proposition 8.6 (The Condition 8.5 Implies the Regularity of $\partial\Omega_u$**)** *Let* $u :$ $B_1 \to \mathbb{R}$ *be a non-negative function such that:*

(a) u is Lipschitz continuous on B_1 *and* $L = \|\nabla u\|_{L^\infty(B_1)}$;
(b) u is non-degenerate in the sense that there is a constant $\eta > 0$ *such that*

$$if \quad y_0 \in \overline{\Omega}_u \cap \partial B_{1/2}, \quad then \quad \|u\|_{L^\infty(B_r(y_0))} \geq \eta r, \quad for\ every \quad r \in (0, 1/2).$$

(c) u satisfies Condition 8.5 for some $\gamma > 0$ *and* $C_1 > 0$.

Then, there is $\rho > 0$ *such that* $\partial\Omega_u$ *is a* $C^{1,\alpha}$ *manifold in* B_ρ, *where* $\alpha := \frac{\gamma}{1+\gamma}$.

Precisely, there are $\rho > 0$ *and a* $C^{1,\alpha}$-*regular function* $g : B'_\rho \to (-\rho, \rho)$ *such that, up to a rotation of the coordinate system of* $\mathbb{R}^d$, *we have*

$$\begin{cases} \left(B'_\rho \times (-\rho, \rho)\right) \cap \Omega_u = \left\{(x', t) \in B'_\rho \times (-\rho, \rho) : g(x') < t\right\}, \\ \left(B'_\rho \times (-\rho, \rho)\right) \setminus \overline{\Omega}_u = \left\{(x', t) \in B'_\rho \times (-\rho, \rho) : g(x') > t\right\}, \\ \left(B'_\rho \times (-\rho, \rho)\right) \cap \partial\Omega_u = \left\{(x', t) \in B'_\rho \times (-\rho, \rho) : g(x') = t\right\}. \end{cases}$$

Lemma 8.7 (Flatness of the Free Boundary $\partial\Omega_u$) *Let $u : B_1 \to \mathbb{R}$ be a non-negative function such that*

(a) u satisfies the Condition 8.5 with constants C_1 and γ.
(b) u is non-degenerate, that is, there is a constant $\eta > 0$ such that

$$\text{if} \quad y_0 \in \overline{\Omega}_u \cap \partial B_{1/2}, \quad \text{then} \quad \|u\|_{L^\infty(B_r(y_0))} \geq \eta r, \quad \text{for every} \quad r \in (0, 1/2).$$

Then, there are constants $C > 0$ and $r_0 > 0$ such that, for every $x_0 \in \partial\Omega_u \cap B_{1/2}$, we have

$$\Omega_{x_0,r} \cap B_1 \supset \{x \in B_1 : x \cdot v_{x_0} > Cr^\gamma\}$$
$$\text{and} \quad \Omega_{x_0,r} \cap \{x \in B_1 : x \cdot v_{x_0} < -Cr^\gamma\} = \emptyset, \tag{8.2}$$

for every $r \in (0, r_0)$, where $\Omega_{x_0,r} := \{u_{x_0,r} > 0\}$.

Proof In order to prove the first part of (8.2), we notice that

$$\|u_{r,x_0} - u_{x_0}\|_{L^\infty(B_1)} \leq C_1 r^\gamma$$

implies that

$$u_{r,x_0}(x) \geq \left(x \cdot v_{x_0} - C_1 r^\gamma\right)_+ \quad \text{for every} \quad x \in B_1.$$

This gives the first inclusion of (8.2) for any constant $C \geq C_1$. In order to prove the second inclusion in (8.2), we suppose that there is a point $y \in B_1$ such that

$$u_{r,x_0}(y) > 0 \quad \text{and} \quad y \cdot v_{x_0} < -Cr^\gamma.$$

This implies that $\tilde{y} := y/2 \in B_{1/2}$ is such that

$$u_{2r,x_0}(\tilde{y}) > 0 \quad \text{and} \quad \tilde{y} \cdot v_{x_0} < -\frac{1}{2}Cr^\gamma.$$

The non-degeneracy of u now implies that

$$\|u_{2r,x_0}\|_{L^\infty(B_\rho(\tilde{y}))} \geq \eta \rho \quad \text{where} \quad \rho := \frac{1}{2}Cr^\gamma.$$

Notice that $u_{x_0} = 0$ on $B_\rho(\tilde{y})$. On the other hand, choosing r_0 such that

$$Cr_0^\gamma \leq 1,$$

we get that $\rho \leq 1/2$ and so $B_\rho(\tilde{y}) \subset B_1$. Thus, we have that

$$\frac{\eta}{2}Cr^\gamma \leq \|u_{2r,x_0} - u_{x_0}\|_{L^\infty(B_1)} \leq C_1(2r)^\gamma,$$

which is a contradiction, if we choose

$$C \geq \frac{2}{\eta}C_1,$$

which concludes the proof by taking

$$C = \left(1 + 2/\eta\right)C_1 \qquad \text{and} \qquad r_0 = \inf\left\{1/2, C^{-\gamma}\right\}.$$

$\square$

Lemma 8.8 (Oscillation of v) *Let $u : B_1 \to \mathbb{R}$ be a Lipschitz continuous function and let $L = \|\nabla u\|_{L^\infty(B_1)}$. Suppose that u satisfies the Condition 8.5 with the constants C_1 and γ. Then, there are constants $R \in (0, 1)$, α and C such that*

$$|v_{x_0} - v_{y_0}| \leq C|x_0 - y_0|^\alpha \qquad \text{for every} \qquad x_0, y_0 \in \partial\Omega_u \cap B_R. \tag{8.3}$$

Precisely, one can take

$$C = 2\sqrt{d+2}\left(L + 2C_1\right), \qquad \alpha = \frac{\gamma}{1+\gamma} \qquad \text{and} \qquad R = 2^{-(2+\gamma)}.$$

Proof Let $\alpha := \frac{\gamma}{1+\gamma}$. Let $x_0, y_0 \in B_R \cap \partial\Omega_u$ and $r := |x_0 - y_0|^{1-\alpha}$. Then, for every $x \in B_1$, we have

$$\left|u_{x_0,r}(x) - u_{y_0,r}(x)\right| = \frac{1}{r}\left|u(x_0 + rx) - u(y_0 + rx)\right| \leq L\frac{|x_0 - y_0|}{r} = L|x_0 - y_0|^\alpha,$$

which gives that

$$\|u_{x_0,r} - u_{y_0,r}\|_{L^\infty(B_1)} \leq L|x_0 - y_0|^\alpha.$$

On the other hand, Condition 8.5 gives that

$$\|u_{x_0,r} - u_{x_0}\|_{L^\infty(B_1)} \leq C_1r^\gamma \qquad \text{and} \qquad \|u_{y_0,r} - u_{y_0}\|_{L^\infty(B_1)} \leq C_1r^\gamma.$$

We notice that in order to apply Condition 8.5 we need that $r \leq 1/2$ and $R \leq 1/2$. We choose R such that $(2R)^{1-\alpha} \leq 1/2$. Thus, by the triangular inequality and the fact that $r^\gamma = |x_0 - y_0|^\alpha$, we obtain

$$\|u_{x_0} - u_{y_0}\|_{L^\infty(B_1)} \leq \left(L + 2C_1\right)|x_0 - y_0|^\alpha.$$

The conclusion now follows by a general argument. Indeed, for any $v_1, v_2 \in \mathbb{R}^d$, we have

$$\left(\frac{\omega_d}{d+2}\right)^{1/2} |v_1 - v_2| = \left(\int_{B_1} |v_1 \cdot x - v_2 \cdot x|^2 \, dx\right)^{1/2}$$

$$\leq \left(\int_{B_1} |(v_1 \cdot x)_+ - (v_2 \cdot x)_+|^2 \, dx\right)^{1/2} + \left(\int_{B_1} |(v_1 \cdot x)_- - (v_2 \cdot x)_-|^2 \, dx\right)^{1/2}$$

$$= 2 \left(\int_{B_1} |(v_1 \cdot x)_+ - (v_2 \cdot x)_+|^2 \, dx\right)^{1/2} \leq 2\omega_d^{1/2} \|(v_1 \cdot x)_+ - (v_2 \cdot x)_+\|_{L_x^\infty(B_1)},$$

which implies that

$$|v_1 - v_2| \leq 2\sqrt{d+2} \, \|(v_1 \cdot x)_+ - (v_2 \cdot x)_+\|_{L_x^\infty(B_1)}.$$

Applying the above estimate to $v_1 = \nu_{x_0}$ and $v_2 = \nu_{y_0}$, we get (8.3). $\square$

Proof of Proposition 8.6 We first notice that, for every $\varepsilon > 0$, there exists $R > 0$ such that, for $x_0 \in \partial\Omega_u \cap B_R$ we have

$$\begin{cases} u > 0 \text{ on } \mathcal{C}_\varepsilon^+(x_0, \nu_{x_0}) \cap B_R(x_0), \\ u = 0 \text{ on } \mathcal{C}_\varepsilon^-(x_0, \nu_{x_0}) \cap B_R(x_0), \end{cases} \tag{8.4}$$

where for a vector $\nu \in \partial B_1$, we denote by $\mathcal{C}_\varepsilon^+(x_0, \nu)$ and $\mathcal{C}_\varepsilon^-(x_0, \nu)$ the cones

$$\mathcal{C}_\varepsilon^\pm(x_0, \nu) := \left\{ x \in \mathbb{R}^d \ : \ \pm \nu \cdot (x - x_0) > \varepsilon |x - x_0| \right\}$$

(see Fig. 8.2).

Indeed, the flatness estimate (8.2) implies (8.4) by taking R such that $CR^\gamma \leq \varepsilon$, where C and γ are the constants from Lemma 8.7.

Let ν_0 be the normal vector at the origin $0 \in \partial\Omega_u$. Without loss of generality we can suppose that $\nu_0 = e_d$. In particular, if $u_0(x) = (x \cdot \nu_0)_+$ is the blow-up limit in

Fig. 8.2 The sets $\mathcal{C}_\varepsilon^\pm(x_0, \nu)$

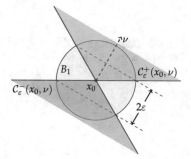

zero, then

$$\Omega_{u_0} = \{u_0 > 0\} = \{(x', x_d) \in \mathbb{R}^{d-1} \times \mathbb{R} : x_d > 0\}.$$

Let $\varepsilon \in (0, 1)$ and $R > 0$ be as in (8.4) and set

$$\rho = R\sqrt{1 - \varepsilon^2} \qquad \text{and} \qquad \ell = \varepsilon R.$$

Let $x' \in B'_\rho$. Then, by (8.4), we have:

- the vertical section

$$\mathcal{S}_+^{x'} := \{(x', t) \in B_R : u(x', t) > 0\}$$

contains the segment

$$\{(x', t) \in B_R : t > \varepsilon R\};$$

- the closed set

$$\mathcal{S}_0^{x'} := \{(x', t) \in B_R : u(x', t) = 0\}$$

contains the segment

$$\{(x', t) \in B_R : t < -\varepsilon R\}.$$

This implies that the function

$$g(x') := \inf\{t \in \mathbb{R} : u(x', T) > 0 \text{ for every } T \in (t, \rho)\},$$

is well defined for $x' \in B'_\rho$ (see Fig. 8.3).
   Let $\delta \leq \rho$. Let $x'_0 \in B'_\delta$ and let $t_0 := g(x'_0)$. By definition, we have

$$x_0 := (x'_0, t_0) \in \partial\Omega_u \cap B_R.$$

Fig. 8.3 Graphicality of the free boundary

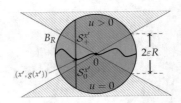

Moreover, by construction, we have

$$-\varepsilon|x_0'| \le g(x_0') \le \varepsilon|x_0'|.$$

Thus,

$$|x_0| \le \delta\sqrt{1+\varepsilon^2} \le \sqrt{2}\,\delta.$$

We next claim that, for δ small enough, we have that

$$u > 0 \text{ on } C_{2\varepsilon}^+(x_0, e_d) \cap B_R(x_0) \qquad \text{and} \qquad u = 0 \text{ on } C_{2\varepsilon}^-(x_0, e_d) \cap B_R(x_0).$$

$$(8.5)$$

Indeed, applying (8.4) for the point x_0, we have

$$u > 0 \text{ on } C_{\varepsilon}^+(x_0, \nu_{x_0}) \cap B_R(x_0) \qquad \text{and} \qquad u = 0 \text{ on } C_{\varepsilon}^-(x_0, \nu_{x_0}) \cap B_R(x_0),$$

so, it is sufficient to prove that

$$C_{2\varepsilon}^\pm(x_0, e_d) \subset C_{\varepsilon}^\pm(x_0, \nu_{x_0}).$$

Let $x \in C_{2\varepsilon}^\pm(x_0, e_d)$. Then,

$$\nu_{x_0} \cdot (x - x_0) = e_d \cdot (x - x_0) + (\nu_{x_0} - e_d) \cdot (x - x_0)$$

$$> 2\varepsilon|x - x_0| - C\big(\sqrt{2}\,\delta\big)^\alpha|x - x_0| > \varepsilon|x - x_0|,$$

where:

- for the first inequality we used the definition of $C_{2\varepsilon}^\pm(x_0, e_d)$ and the following estimate, which is a consequence of Lemma 8.8:

$$|\nu_{x_0} - e_d| \le C|x_0|^\alpha \le C\big(\sqrt{2}\,\delta\big)^\alpha\,;$$

- for the second in equality, we choose δ such that $C\big(\sqrt{2}\,\delta\big)^\alpha \le \varepsilon.$

This proves (8.5). As a consequence, we obtain that the sections $S_+^{x'}$ and $S_0^{x'}$ are segments:

$$\begin{cases} \big(B_\delta' \times (-\delta, \delta)\big) \cap \Omega_u = \big\{(x', t) \in B_\delta' \times (-\delta, \delta) : g(x') < t\big\}, \\ \big(B_\delta' \times (-\delta, \delta)\big) \setminus \Omega_u = \big\{(x', t) \in B_\delta' \times (-\delta, \delta) : g(x') \ge t\big\}, \end{cases}$$

and so, the free boundary is precisely the graph of g, that is,

$$\big(B_\delta' \times (-\delta, \delta)\big) \cap \partial\Omega_u = \big\{(x', t) \in B_\delta' \times (-\delta, \delta) : g(x') = t\big\}.$$

We next prove that the function $g : B'_\delta \to \mathbb{R}$ is Lipschitz continuous on B'_δ. Also this follows by the uniform cone condition (8.5). Indeed, let

$$x'_1, x'_2 \in B'_\delta, \quad x_1 = (x'_1, g(x'_1)) \quad \text{and} \quad x_2 = (x'_2, g(x'_2)).$$

Since $x_1 \notin C^+_{2\varepsilon}(x_2, e_d)$, we have that

$$g(x'_1) - g(x'_2) = (x_1 - x_2) \cdot e_d \le 2\varepsilon|x_1 - x_2| \le 2\varepsilon|x'_1 - x'_2| + 2\varepsilon|g(x'_1) - g(x'_2)|.$$

Analogously, $x_2 \notin C^+_{2\varepsilon}(x_1, e_d)$ implies that

$$g(x'_1) - g(x'_2) \le 2\varepsilon|x'_1 - x'_2| + 2\varepsilon|g(x'_1) - g(x'_2)|,$$

and the two estimates give

$$(1 - 2\varepsilon)\,|g(x'_1) - g(x'_2)| \le 2\varepsilon|x'_1 - x'_2|,$$

and finally, choosing $\varepsilon \le 1/4$, we get

$$|g(x'_1) - g(x'_2)| \le 4\varepsilon|x'_1 - x'_2|,$$

which concludes the proof of the Lipschitz continuity of g.

We will next show that g is differentiable. Indeed, let $x'_0 \in B'_\delta$. Now, the improvement of flatness at $x_0 = (x'_0, g(x'_0))$ implies that

$$-C|x - x_0|^{1+\gamma} \le (x - x_0) \cdot \nu_{x_0} \le C|x - x_0|^{1+\gamma},$$

for any $x = (x', g(x'))$ with $x' \in B'_\delta$. For the sake of simplicity, we set $\nu := \nu_{x_0}$ and $\nu = (\nu', \nu_d) \in \mathbb{R}^{d-1} \times \mathbb{R}$. Since

$$(x - x_0) \cdot \nu_{x_0} = (x' - x'_0) \cdot \nu' + \big(g(x') - g(x'_0)\big)\nu_d,$$

we get that

$$\left| g(x') - g(x'_0) - (x' - x'_0) \cdot \frac{\nu'}{\nu_d} \right| \le \frac{C}{\nu_d}(1 + \varepsilon)^{1+\gamma}|x' - x'_0|^{1+\gamma}.$$

This implies that g is differentiable at x'_0 and that $\nabla g(x'_0) = \frac{\nu'}{\nu_d}$. Finally, the α-Hölder continuity of $\nabla g : B'_\delta \to \mathbb{R}^{d-1}$ follows by the γ-Hölder continuity of the map $x \mapsto \nu_x$. Indeed, for any $x', y' \in B'_\delta$, $x = (x', g(x'))$ and $y = (y', g(y'))$ we have that

$$|\nu_x - \nu_y| \le |x - y|^\alpha \le (1 + \varepsilon)^\alpha |x' - y'|^\alpha,$$

which implies the Hölder continuity of all the components of the map $B'_\delta \ni x \mapsto \nu_x \in \mathbb{R}^d$ and thus, of the gradient ∇g. This concludes the proof of Proposition 8.6.

$\square$

Chapter 9
The Weiss Monotonicity Formula and Its Consequences

This chapter is dedicated to the monotonicity formula for the boundary adjusted energy introduced by Weiss in [52]. Precisely, for every $\Lambda \geq 0$ and every $u \in H^1(B_1)$ we define

$$W_\Lambda(u) := \int_{B_1} |\nabla u|^2 \, dx - \int_{\partial B_1} u^2 \, d\mathcal{H}^{d-1} + \Lambda |\Omega_u \cap B_1|,$$

where we recall that $\Omega_u := \{u > 0\}$. In particular, we have

$$W_0(u) = \int_{B_1} |\nabla u|^2 \, dx - \int_{\partial B_1} u^2 \, d\mathcal{H}^{d-1} \qquad \text{and} \qquad W_\Lambda(u) = W_0(u) + \Lambda |\Omega_u \cap B_1|.$$

This chapter is organized as follows:

In Sect. 9.1 we prove several preliminary results for the Weiss' boundary adjusted energy, which hold for a general Sobolev function u defined on an open set $D \subset \mathbb{R}^d$. In particular, in Lemma 9.1 we prove that the function $(x_0, r) \mapsto W_\Lambda(u_{x_0,r})$ is continuous (where it is defined), where we recall that $u_{x_0,r}(x) := \frac{1}{r} u(x_0 + rx)$; in Lemma 9.2, we compute the derivative of $W_\Lambda(u_{x_0,r})$ with respect to r and we prove that

$$\frac{\partial}{\partial r} W_\Lambda(u_{x_0,r}) = \frac{d}{r} \big(W_\Lambda(z_{x_0,r}) - W_\Lambda(u_{x_0,r}) \big) + \frac{1}{r} \mathcal{D}(u_{x_0,r}),$$

where $z_{x_0,r}$ is the one-homogeneous extension defined in Lemma 9.2 while the *deviation* $\mathcal{D}(u_{x_0,r})$ is defined as

$$\mathcal{D}(u_{x_0,r}) := \int_{\partial B_1} |x \cdot \nabla u_{x_0,r} - u_{x_0,r}|^2 \, d\mathcal{H}^{d-1},$$

© The Author(s) 2023
B. Velichkov, *Regularity of the One-phase Free Boundaries*,
Lecture Notes of the Unione Matematica Italiana 28,
https://doi.org/10.1007/978-3-031-13238-4_9

and measures at what extent the function is not one-homogeneous (see Lemma 9.3) and controls the oscillation of u from scale to scale, which is measured by the norm $\|u_{x_0,r} - u_{x_0,s}\|_{L^2(\partial B_1)}$. Finally, in Proposition 9.4, as a direct consequence of the Weiss formula (Lemma 9.2), we obtain that, if u is a (local) minimizer of $\mathcal{F}_\Lambda$ in D, then the Weiss energy $W(u_{x_0,r})$ is monotone increasing in r.

In Sect. 9.2 we introduce the notion of *stationary free boundary*, that is, the free boundary $\partial \Omega_u \cap D$ of a function $u : D \to \mathbb{R}$, which is stationary for the functional $\mathcal{F}_\Lambda$ with respect to internal perturbations with vector fields compactly supported in D. In Lemma 9.5, we compute the variation of the energy $\mathcal{F}_\Lambda$ with respect to a compactly supported vector field $\xi \in C_c^\infty(D; \mathbb{R}^d)$, which is simply defined as

$$\delta \mathcal{F}_\Lambda(u, D)[\xi] := \frac{\partial}{\partial t}\Big|_{t=0} \mathcal{F}_\Lambda(u_t, D),$$

where $u_t : D \to \mathbb{R}$ is defined through the identity $u_t(x + t\xi(x)) = u(x)$. We say that a function is stationary (see Definition 9.7), if the first variation is zero with respect to any vector field, that is, if

$$\delta \mathcal{F}_\Lambda(u, D)[\xi] = 0 \qquad \text{for every} \qquad \xi \in C_c^\infty(D; \mathbb{R}^d).$$

In Lemma 9.6 we show that if u is a minimizer of $\mathcal{F}_\Lambda$ in D, then it is stationary in D. Then, in Lemma 9.8, we prove that every stationary function satisfies an *equipartition-of-the-energy* identity; in Lemma 9.9, we prove that the equipartition of the energy is sufficient for the monotonicity of the Weiss energy. In particular, the monotonicity formula holds for stationary free boundaries. The result of Sect. 9.2 are fundamental for the proof of Theorem 1.9, but we do not need them in the proof of Theorem 1.4, where we can use directly Proposition 9.4.

In Sect. 9.3 we give the sufficient conditions for the homogeneity of the blow-up limits of a function $u : D \to \mathbb{R}$ (Lemma 9.10). We then apply this result to minimizers of $\mathcal{F}_\Lambda$ (Proposition 9.12), but we will also use it in the context of Theorem 1.9. This is why the exposition contains the intermediate Lemma 9.11.

In Sect. 9.4 we prove that the only one-homogeneous global solutions in dimension two are the half-plane solutions (see Proposition 9.13). In particular, this means that $d^* \geq 3$.

In Sect. 9.5 we give another proof of the fact that the minimizers of $\mathcal{F}_\Lambda$ are viscosity solutions (Proposition 7.1). Our main result is Proposition 9.18, which applies to minimizers of $\mathcal{F}_\Lambda$, but also in the context of Theorem 1.9.

Finally, in Sect. 9.6, we use the Weiss monotonicity formula to relate the energy density

$$\lim_{r \to 0} W(u_{x_0,r}),$$

of a minimizer u of $\mathcal{F}_\Lambda$, to the Lebesgue density

$$\lim_{r \to 0} \frac{|\Omega_u \cap B_r(x_0)|}{|B_r|},$$

of the set Ω_u, at every point of the free boundary $x_0 \in \partial\Omega_u$ (Lemma 9.20). Moreover, we characterize the regular part of the free boundary $Reg(\partial\Omega_u)$ in terms of the energy and the Lebesgue densities (Lemma 9.22). We will not use the results from Sect. 9.6 in the proofs of Theorems 1.2, 1.4, 1.9 and 1.10, but they remain an interesting application of the monotonicity formula and the homogeneity of the blow-up limits and were used, for instance, in the analysis of the vectorial free boundaries (see [41]).

9.1 The Weiss Boundary Adjusted Energy

Let $u \in H^1(B_r(x_0))$ be a given function on the ball $B_r(x_0) \subset \mathbb{R}^d$ and consider the rescaling

$$u_{r,x_0} \in H^1(B_1) \qquad \text{where} \qquad u_{r,x_0}(x) = \frac{1}{r}u(x_0 + rx).$$

We notice that the different terms of the energy W_Λ have the following scaling properties:

$$\int_{B_1} |\nabla u_{r,x_0}|^2 \, dx = \frac{1}{r^d} \int_{B_r(x_0)} |\nabla u|^2 \, dx \,,$$

$$\int_{\partial B_1} u_{r,x_0}^2 \, d\mathcal{H}^{d-1} = \frac{1}{r^{d+1}} \int_{\partial B_r(x_0)} u^2 \, d\mathcal{H}^{d-1}$$

$$\text{and} \qquad \left|\{u_{x_0,r} > 0\} \cap B_1\right| = \frac{1}{r^d}\left|\{u > 0\} \cap B_r(x_0)\right|.$$

Thus, we have

$$W_\Lambda(u_{x_0,r}) = \frac{1}{r^d} \int_{B_r(x_0)} |\nabla u|^2 \, dx - \frac{1}{r^{d+1}} \int_{\partial B_r(x_0)} u^2 \, d\mathcal{H}^{d-1} + \frac{\Lambda}{r^d}\left|\{u > 0\} \cap B_r(x_0)\right|.$$

In particular, since u is a Sobolev function, the function $(x_0, r) \mapsto W_\Lambda(u_{x_0,r})$ is continuous, where it is defined. We give the precise statement in the following lemma.

Lemma 9.1 (Continuity of the Function $(x_0, r) \mapsto W_\Lambda(u_{x_0,r})$) *Let D be a bounded open set in $\mathbb{R}^d$ and let $u \in H^1(D)$. Let $\delta > 0$ and let D_δ be the set*

$$D_\delta := \left\{x \in D \ : \ \mathrm{dist}\,(x, \partial D) < \delta\right\}.$$

Then, the function

$$\Phi_u : D_\delta \times (0, \delta) \to \mathbb{R}, \qquad \Phi_u(x_0, r) := W_\Lambda(u_{x_0, r}),$$

is continuous.

Proof The continuity of the terms

$$(x_0, r) \mapsto \frac{1}{r^d} \int_{B_r(x_0)} |\nabla u|^2 \, dx \qquad \text{and} \qquad (x_0, r) \mapsto \frac{1}{r^d} |\{u > 0\} \cap B_r(x_0)|,$$

follows by the fact that if $f : D \to \mathbb{R}$ is a function in $L^1(D)$, then the map

$$(x_0, r) \mapsto \int_{B_r(x_0)} f(x) \, dx,$$

is continuous, which in turn follows by the dominated convergence theorem. In order to prove the continuity of the function

$$(x_0, r) \mapsto \frac{1}{r^{d+1}} \int_{\partial B_r(x_0)} u^2 \, d\mathcal{H}^{d-1},$$

we consider the sequence to $(x_n, r_n) \in D_\delta \times (0, \delta)$ converging to a point $(x_0, r_0) \in D_\delta \times (0, \delta)$. We first notice that reasoning as above, we have

$$\lim_{n \to \infty} \|\nabla u_{x_n, r_n}\|_{L^2(B_1)} = \|\nabla u_{x_0, r_0}\|_{L^2(B_1)} \qquad \text{and}$$

$$\lim_{n \to \infty} \|u_{x_n, r_n}\|_{L^2(B_1)} = \|u_{x_0, r_0}\|_{L^2(B_1)}.$$

Next, we notice that u_{x_n, r_n} converges weakly in $H^1(B_1)$ to u_{x_0, r_0}. In fact, for any $\phi \in C_c^\infty(B_1)$ we have

$$\lim_{n \to \infty} \int_{B_1} \nabla \phi \cdot \nabla u_{x_n, r_n} \, dx = \lim_{n \to \infty} \int_{B_1} \nabla \phi(x) \cdot \nabla u(x_n + r_n x) \, dx$$

$$= \lim_{n \to \infty} \int_{B_1} \nabla \phi\left(\frac{y - x_n}{r_n}\right) \cdot \nabla u(y) \, dy$$

$$= \int_{B_1} \nabla \phi\left(\frac{y - x_0}{r_0}\right) \cdot \nabla u(y) \, dy$$

$$= \int_{B_1} \nabla \phi \cdot \nabla u_{x_0, r_0} \, dx.$$

Now, since the norm of u_{x_n, r_n} converges to the norm of u_{x_0, r_0}, we get that

$$u_{x_n, r_n} \to u_{x_0, r_0} \qquad \text{strongly in} \qquad H^1(B_1).$$

By the trace inequality, we have that

$$u_{x_n,r_n} \to u_{x_0,r_0} \qquad \text{strongly in} \qquad L^2(\partial B_1),$$

which concludes the proof. □

Lemma 9.2 (Derivative of the Weiss' Energy) *Let D be a bounded open set in $\mathbb{R}^d$ and let $u \in H^1(D)$. Let $x_0 \in D$ and $\delta = \text{dist}(x_0, \partial D)$. Then, the function $\Phi_u(x_0, \cdot)$ is differentiable almost everywhere on $(0, \delta)$ and for (almost) every $r \in (0, \delta)$, we have*

$$\frac{\partial}{\partial r} W_\Lambda(u_{x_0,r}) = \frac{d}{r}\left(W_\Lambda(z_{x_0,r}) - W_\Lambda(u_{x_0,r})\right)$$

$$+ \frac{1}{r}\int_{\partial B_1} |x \cdot \nabla u_{x_0,r} - u_{x_0,r}|^2 \, d\mathcal{H}^{d-1}, \qquad (9.1)$$

where $z_{x_0,r} : B_1 \to \mathbb{R}$ is the one-homogeneous extension of $u_{x_0,r}$ in B_1:

$$z_{x_0,r}(x) := |x|\, u_{x_0,r}(x/|x|).$$

Proof Without loss of generality we can assume $x_0 = 0$. We recall that $u_r := u_{0,r}$.

We first notice that the function $r \mapsto |\Omega_u \cap B_r|$ is differentiable almost everywhere and that for almost every $r \in (0, \delta)$ we have

$$\frac{\partial}{\partial r}\left(\frac{1}{r^d}|\Omega_u \cap B_r|\right) = -\frac{d}{r^{d+1}}|\Omega_u \cap B_r| + \frac{1}{r^d}\mathcal{H}^{d-1}(\Omega_u \cap \partial B_r),$$

which can be written as

$$\frac{\partial}{\partial r}\left(\frac{1}{r^d}|\Omega_u \cap B_r|\right) = -\frac{d}{r}|\Omega_{u_r} \cap B_1| + \frac{d}{r}|\Omega_{z_r} \cap B_1|. \qquad (9.2)$$

In fact, we have

$$|\Omega_{z_r} \cap B_1| = \int_0^1 \mathcal{H}^{d-1}(\Omega_{u_r} \cap \partial B_1)s^{d-1}\, ds = \frac{1}{d}\mathcal{H}^{d-1}(\Omega_{u_r} \cap \partial B_1) = \frac{r^{d-1}}{d}\mathcal{H}^{d-1}(\Omega_u \cap \partial B_r).$$

Thus, (9.2) implies that it is sufficient to prove (9.1) in the case $\Lambda = 0$.

As above, we notice that the function $r \mapsto \int_{B_r} |\nabla u|^2 \, dx$ is differentiable almost-everywhere and that we have

$$\frac{\partial}{\partial r}\left(\frac{1}{r^d}\int_{B_r}|\nabla u|^2\, dx\right) = -\frac{d}{r^{d+1}}\int_{B_r}|\nabla u|^2\, dx + \frac{1}{r^d}\int_{\partial B_r}|\nabla u|^2\, d\mathcal{H}^{d-1}$$

$$= -\frac{d}{r^{d+1}}\int_{B_r}|\nabla u|^2\, dx + \frac{1}{r}\int_{\partial B_1}|\nabla u_r|^2\, d\mathcal{H}^{d-1}. \qquad (9.3)$$

In order to deal with the boundary term, we first compute

$$\frac{\partial}{\partial r} \left(\frac{1}{r^{d-1}} \int_{\partial B_r} u^2(x)\, d\mathcal{H}^{d-1}(x) \right) = \frac{\partial}{\partial r} \int_{\partial B_1} u(ry)^2\, d\mathcal{H}^{d-1}(y)$$

$$= 2 \int_{\partial B_1} u(ry)\, y \cdot \nabla u(ry)\, d\mathcal{H}^{d-1}(y)$$

$$= 2r \int_{\partial B_1} u_r\, (x \cdot \nabla u_r)\, d\mathcal{H}^{d-1}(x)$$

Thus, we have

$$\frac{\partial}{\partial r} \left(\frac{1}{r^{d+1}} \int_{\partial B_r} u^2\, d\mathcal{H}^{d-1} \right) = -\frac{2}{r^{d+2}} \int_{\partial B_r} u^2\, d\mathcal{H}^{d-1} + \frac{2}{r} \int_{\partial B_1} u_r\, (x \cdot \nabla u_r)\, d\mathcal{H}^{d-1}.$$

$$(9.4)$$

Now, we notice that for every r such that $u_r = z_r \in H^1(\partial B_1)$, we can write the function $z_r : B_1 \to \mathbb{R}$ in polar coordinates $\rho \in (0,1]$, $\theta \in \mathbb{S}^{d-1}$ as $z_r(\rho, \theta) = \rho\, z_r(1, \theta)$ and we have

$$W_0(z_r) = \int_{B_1} |\nabla z_r|^2\, dx - \int_{\partial B_1} z_r^2\, d\mathcal{H}^{d-1}$$

$$= \int_0^1 r^{d-1}\, dr \int_{\mathbb{S}^{d-1}} \left(z_r^2(1, \theta) + |\nabla_\theta z_r|^2 \right) d\theta - \int_{\mathbb{S}^{d-1}} z_r^2(1, \theta)\, d\theta$$

$$= \frac{1}{d} \int_{\mathbb{S}^{d-1}} |\nabla_\theta z_r|^2\, d\theta - \frac{d-1}{d} \int_{\mathbb{S}^{d-1}} z_r^2(1, \theta)\, d\theta$$

$$= \frac{1}{d} \int_{\partial B_1} \left(|\nabla u_r|^2 - (x \cdot \nabla u_r)^2 \right) d\mathcal{H}^{d-1} - \frac{d-1}{d} \int_{\partial B_1} u_r^2\, d\mathcal{H}^{d-1}.$$

$$(9.5)$$

Now, putting together (9.3), (9.4) and (9.5), we get that

$$\frac{\partial}{\partial r} W_0(u_{x_0, r}) = \frac{d}{r} \left(W_0(z_{x_0, r}) - W_0(u_{x_0, r}) \right) + \frac{1}{r} \int_{\partial B_1} |x \cdot \nabla u_{x_0, r} - u_{x_0, r}|^2\, d\mathcal{H}^{d-1},$$

which concludes the proof. $\square$

We now define the deviation $\mathcal{D}$ as

$$\mathcal{D}(\phi) := \int_{\partial B_1} |x \cdot \nabla \phi - \phi|^2\, d\mathcal{H}^{d-1}.$$

Thus, (9.1) can be written as

$$\frac{\partial}{\partial r} W_\Lambda(u_{x_0,r}) = \frac{d}{r}\left(W_\Lambda(z_{x_0,r}) - W_\Lambda(u_{x_0,r})\right) + \frac{1}{r}\mathcal{D}(u_{x_0,r}).$$

In the next lemma we show that the deviation $\mathcal{D}(u_{x_0,r})$ controls the oscillation of u.

Lemma 9.3 (The Deviation Controls the Oscillation of the Blow-Up Sequence)
Let D be a bounded open set in $\mathbb{R}^d$ and let $u \in H^1(D)$. Let $x_0 \in D$ and $\delta = dist(x_0, \partial D)$. Then, for almost every $0 < r < R < \delta$, we have

$$\|u_{x_0,R} - u_{x_0,r}\|^2_{L^2(\partial B_1)} \leq \frac{1}{r}\int_r^R \mathcal{D}(u_{x_0,s})\, ds.$$

In particular, if $\mathcal{D}(u_{x_0,s}) = 0$ for every $s \in (0,\delta)$, then the function $u_{x_0,\delta} : B_1 \to \mathbb{R}$ is one-homogeneous, that is

$$u(x_0 + rx) = ru(x_0 + x) \quad \text{for every} \quad |x| \leq \delta \quad \text{and every} \quad r \leq 1.$$

Proof We set for simplicity, $x_0 = 0$ and $u_r := u_{x_0,r}$. For any $x \in \partial B_1$, we have

$$\frac{u(Rx)}{R} - \frac{u(rx)}{r} = \int_r^R \left(\frac{x \cdot (\nabla u)(sx)}{s} - \frac{u(sx)}{s^2}\right) ds = \int_r^R \frac{1}{s}\left(x \cdot \nabla u_s(x) - u_s(x)\right) ds.$$

Integrating over the sphere ∂B_1 and using the Cauchy-Schwarz inequality, we obtain

$$\int_{\partial B_1} |u_R - u_r|^2\, d\mathcal{H}^{d-1} \leq \int_{\partial B_1} \left(\int_r^R \frac{1}{s}|x \cdot \nabla u_s - u_s|\, ds\right)^2 d\mathcal{H}^{d-1}$$

$$\leq \int_{\partial B_1} \left(\int_r^R s^{-2}ds\right)\left(\int_r^R |x \cdot \nabla u_s - u_s|^2 ds\right) d\mathcal{H}^{d-1}$$

$$= \left(\frac{1}{r} - \frac{1}{R}\right)\int_r^R \mathcal{D}(u_s)\, ds.$$

which concludes the proof.   □

We conclude this subsection with the following proposition.

Proposition 9.4 (Weiss Monotonicity Formula) *Let D be a bounded open set in $\mathbb{R}^d$ and let $u \in H^1(D)$ be a minimizer of $\mathcal{F}_\Lambda$ in D. Let $x_0 \in D$ and $\delta_{x_0} = dist(x_0, \partial D)$. Then the function $r \mapsto W_\Lambda(u_{x_0,r})$ is non-decreasing on the interval $(0, \delta_{x_0})$.*

Proof By Lemma 9.2 we have that

$$\frac{\partial}{\partial r} W_\Lambda(u_{x_0,r}) \geq \frac{d}{r}\big(W_\Lambda(z_{x_0,r}) - W_\Lambda(u_{x_0,r})\big).$$

Now, since $u_{x_0,r}$ is a minimizer of $\mathcal{F}_\Lambda$ in B_1 and since by definition $z_{x_0,r} = u_{x_0,r}$ on ∂B_1, we get that $\frac{\partial}{\partial r} W_\Lambda(u_{x_0,r}) \geq 0$, which concludes the proof.   □

9.2 Stationary Free Boundaries

In this section we introduce the notion of a stationary free boundary (Definition 9.7) and we prove a monotonicity formula for the Weiss energy (Proposition 9.9).

Lemma 9.5 (First Variation of the Energy) *Suppose that $D \subset \mathbb{R}^d$ is a bounded open set and that $u \in H^1(D)$. Let $\xi \in C_c^\infty(D; \mathbb{R}^d)$ be a given vector field with compact support in D and let Ψ_t be the diffeomorphism*

$$\Psi_t(x) = x + t\xi(x) \quad \text{for every} \quad x \in D.$$

Then,

(i) for t small enough, $\Psi_t : D \to D$ is a diffeomorphism and setting $\Phi_t := \Psi_t^{-1}$, the function $u_t := u \circ \Phi_t$ is well-defined and belongs to $H^1(D)$;

(ii) the function $t \mapsto \displaystyle\int_D |\nabla u_t|^2 \, dx$ is differentiable at $t = 0$ and

$$\frac{\partial}{\partial t}\Big|_{t=0} \int_D |\nabla u_t|^2 \, dx = \int_D \Big(-2\nabla u \cdot D\xi \nabla u + |\nabla u|^2 \mathrm{div}\,\xi\Big) \, dx;$$

(iii) the function $t \mapsto |\Omega_{u_t} \cap D|$ is differentiable at $t = 0$ and

$$\frac{\partial}{\partial t}\Big|_{t=0} |\Omega_{u_t} \cap D| = \int_{\Omega_u \cap D} \mathrm{div}\,\xi \, dx.$$

(iv) if Ω_u is open, if $\partial\Omega_u$ is a C^2 regular in D and if $u \in C^2(\Omega_u)$, then

$$\frac{\partial}{\partial t}\Big|_{t=0} \int_D |\nabla u_t|^2 \, dx = -\int_{\partial\Omega_u} \xi \cdot \nu \, |\nabla u|^2 \, d\mathcal{H}^{d-1} \qquad and$$

$$\frac{\partial}{\partial t}\Big|_{t=0} |\Omega_{u_t} \cap D| = \int_{\partial\Omega_u} \xi \cdot \nu \, d\mathcal{H}^{d-1},$$

where $\nu(x)$ is the exterior normal to $\partial\Omega$ at the point $x \in \partial\Omega$.

Proof The first claim follows by the fact that ξ is smooth and compactly supported in D. Thus, we start directly by proving (ii). We use the conventions

$$
x = \begin{pmatrix} x_1 \\ \vdots \\ x_d \end{pmatrix}, \quad
\nabla u = \begin{pmatrix} \partial_1 u \\ \vdots \\ \partial_d u \end{pmatrix}, \quad
\Phi = \begin{pmatrix} \Phi_1 \\ \vdots \\ \Phi_d \end{pmatrix}, \quad
D\Phi = \begin{pmatrix} \partial_1 \Phi_1 & \cdots & \partial_1 \Phi_d \\ \vdots & \ddots & \vdots \\ \partial_d \Phi_1 & \cdots & \partial_d \Phi_d \end{pmatrix},
$$

for general $u : \mathbb{R}^d \to \mathbb{R}$ and $\Phi : \mathbb{R}^d \to \mathbb{R}^d$, so that

$$
\nabla(u \circ \Phi)(x) = D\Phi(x)\nabla u(\Phi(x)).
$$

In our case $u_t = u \circ \Phi_t$, by the change of variables $y = \Phi_t(x)$ (thus, $x = \Psi_t(y)$), we get

$$
\int_D |\nabla u_t|^2(x)\, dx = \int_D \Big(D\Phi_t(\Psi_t(y))\nabla u(y)\Big) \cdot \Big(D\Phi_t(\Psi_t(y))\nabla u(y)\Big) |\det D\Psi_t(y)|\, dy
$$

$$
= \int_D \nabla u(y) \cdot \Big([D\Phi_t(\Psi_t(y))]^T D\Phi_t(\Psi_t(y))\Big)\nabla u(y) |\det D\Psi_t(y)|\, dy
$$

$$
= \int_D \nabla u(y) \cdot \Big([D\Psi_t(y)]^{-T}[D\Psi_t(y)]^{-1}\Big)\nabla u(y) |\det D\Psi_t(y)|\, dy
$$

We now notice that

$$
D\Psi_t = Id + t D\xi, \qquad [D\Psi_t]^{-1} = Id - t D\xi + o(t), \qquad \det D\Psi_t = 1 + t \operatorname{div}\xi + o(t),
$$

and we calculate

$$
\int_D |\nabla u_t|^2\, dx = \int_D |\nabla u|^2\, dx + t \int_D \Big(|\nabla u|^2 \operatorname{div}\xi - 2\nabla u \cdot D\xi\, \nabla u\Big)dx + o(t),
$$

which concludes the proof of (ii).

In order to prove (iii), we notice that

$$
x \in \Omega_{u_t} \iff u_t(x) > 0 \iff \Phi_t(x) \in \Omega_u.
$$

This means that $\mathbb{1}_{\Omega_{u_t}} = \mathbb{1}_{\Omega_u} \circ \Phi_t$, and so, we can compute

$$
|\Omega_{u_t}| = \int_D \mathbb{1}_{\Omega_u}(\Phi_t(x))\, dx = \int_D \mathbb{1}_{\Omega_u}(y)|\det D\Psi_t(y)|\, dy
$$

$$
= \int_{\Omega_u} \big(1 + t \operatorname{div}\xi(y) + o(t)\big)\, dy = |\Omega_u| + t \int_{\Omega_u} \operatorname{div}\xi\, dx + o(t),
$$

which proves (iii).

We now prove (iv). Assume that u is C^2 in the open set Ω_u. Then, setting $\xi = (\xi_1, \ldots, \xi_d)$ and using the convention for summation over the repeating indices, we compute

$$
\begin{aligned}
|\nabla u|^2 \mathrm{div}\, \xi - 2\nabla u\, D\xi \cdot \nabla u &= \partial_i u\, \partial_i u\, \partial_j\, \xi_j - 2\partial_i u\, \partial_j \xi_i\, \partial_j u \\
&= \partial_i u\, \partial_i u\, \partial_j \xi_j - 2\partial_j(\partial_i u\, \xi_i\, \partial_j u) + 2\partial_{ij} u\, \xi_i \partial_j u \\
&\quad + 2\partial_i \xi_i\, \partial_{jj} u \\
&= \partial_i u \partial_i u \partial_j \xi_j - 2\partial_j(\partial_i u\, \xi_i\, \partial_j u) + 2\partial_{ij} u\, \xi_i\, \partial_j u \\
&= \partial_i u\, \partial_i u\, \partial_j \xi_j - 2\partial_j(\partial_i u\, \xi_i\, \partial_j u) + \partial_i(\partial_j u\, \xi_i\, \partial_j u) \\
&\quad - \partial_j u\, \partial_i \xi_i\, \partial_j u \\
&= -2\partial_j(\partial_i u\, \xi_i\, \partial_j u) + \partial_j(\partial_i u\, \xi_j\, \partial_i u) \\
&= \mathrm{div}\left(|\nabla u|^2 \xi - 2(\xi \cdot \nabla u)\nabla u\right).
\end{aligned}
$$

Integrating by parts we obtain

$$
\int_{\Omega_u} \mathrm{div}\left(|\nabla u|^2 \xi - 2(\xi \cdot \nabla u)\nabla u\right) dx
$$

$$
= \int_{\partial\Omega_u}\left(|\nabla u|^2(\xi \cdot \nu) - 2(\xi \cdot \nabla u)(\nabla u \cdot \nu)\right) d\mathcal{H}^{d-1}.
$$

Since $u = 0$ on $\partial\Omega_u$ and positive in Ω_u, we have that $\nabla u = \nu|\nabla u|$. Thus,

$$
\int_{\Omega_u} \mathrm{div}\left(|\nabla u|^2 \xi - 2(\xi \cdot \nabla u)\nabla u\right) dx = -\int_{\partial\Omega_u} |\nabla u|^2(\xi \cdot \nu)\, d\mathcal{H}^{d-1},
$$

which proves the first part of the claim (iv). The second part of (iv) follows by a simple integration by parts in Ω_u. □

As a consequence of Lemma 9.5 we obtain that for every $\Lambda \in \mathbb{R}$, $u \in H^1(D)$ and vector field $\xi \in C_c^\infty(D; \mathbb{R}^d)$ we can define *the first variation of $\mathcal{F}_\Lambda$ at u in the direction ξ* as

$$
\delta\mathcal{F}_\Lambda(u, D)[\xi] := \int_D \left(-2\nabla u \cdot D\xi\, \nabla u + |\nabla u|^2 \mathrm{div}\, \xi + \Lambda\, \mathbb{1}_{\Omega_u} \mathrm{div}\, \xi\right) dx. \tag{9.6}
$$

Lemma 9.6 (The Minimizers have Zero First Variation) *Let D be a bounded open set in $\mathbb{R}^d$ and let $u \in H^1(D)$ be a minimizer of $\mathcal{F}_\Lambda$ in D. Then,*

$$
\delta\mathcal{F}_\Lambda(u, D)[\xi] = 0 \qquad \text{for every vector field} \qquad \xi \in C_c^\infty(D; \mathbb{R}^d).
$$

If, moreover, $\partial\Omega_u$ is C^2 smooth in D, then

$$
|\nabla u| = \sqrt{\Lambda} \qquad on \qquad \partial\Omega_u \cap D. \tag{9.7}
$$

Proof The first part of the statement follows directly by Lemma 9.5. In order to prove the second part, we notice that in the case when $\partial\Omega_u$ is smooth, we have

$$\delta\mathcal{F}_\Lambda(u, D)[\xi] = \int_{\partial\Omega_u} \left(\Lambda - |\nabla u|^2\right) \xi \cdot \nu \, d\mathcal{H}^{d-1},$$

for every vector field $\xi \in C_c^\infty(D; \mathbb{R}^d)$. This implies (9.7). $\qquad\square$

Definition 9.7 (Stationary Free Boundaries) Let $D \subset \mathbb{R}^d$ be a bounded open set and $u \in H^1(D)$ be a non-negative function such that

$$\delta\mathcal{F}_\Lambda(u, D)[\xi] = 0 \qquad \text{for every vector field} \qquad \xi \in C_c^\infty(D; \mathbb{R}^d).$$

Then, we say that the function u and the free boundary $\partial\Omega_u$ are stationary for $\mathcal{F}_\Lambda$.

As a consequence of Lemma 9.6 we obtain the following.

Lemma 9.8 (Equipartition of the Energy) *Suppose that D is a bounded open set in $\mathbb{R}^d$ and $u \in H^1(D)$ is a non-negative function which is stationary for $\mathcal{F}_\Lambda$ (in the sense of Definition 9.7). Then, for every $x_0 \in D$ and every $0 < r < \mathrm{dist}(x_0, \partial D)$, we have*

$$W_\Lambda(z_{x_0,r}) - W_\Lambda(u_{x_0,r}) = \frac{1}{d} \int_{\partial B_1} |x \cdot \nabla u_{x_0,r} - u_{x_0,r}|^2 \, d\mathcal{H}^{d-1}, \qquad (9.8)$$

where we recall that $u_{x_0,r}(x) = \frac{1}{r}u(x_0 + rx)$ and that $z_{x_0,r}$ is the one-homogeneous extension of $u_{x_0,r}$ in B_1, that is, $z_{x_0,r}(x) = |x|u_{x_0,r}(x/|x|)$.

Proof Without loss of generality, we assume that $x_0 = 0$. For every $\varepsilon > 0$, we consider a function $\phi_\varepsilon \in C_c^\infty(B_r)$ such that

$$\phi_\varepsilon = 1 \quad \text{in} \quad B_{(1-\varepsilon)r}, \qquad \nabla\phi_\varepsilon(x) = -\frac{1}{r\varepsilon}\frac{x}{|x|} + o(\varepsilon) \quad \text{in} \quad B_r \setminus B_{(1-\varepsilon)r}.$$

Taking the vector field $\xi_\varepsilon(x) = x\phi_\varepsilon(x)$ we get that

$$\mathrm{div}\,\xi_\varepsilon(x) = d\phi_\varepsilon(x) + x \cdot \nabla\phi_\varepsilon(x),$$

$$D\xi_\varepsilon(x) = \phi_\varepsilon(x)Id + x \otimes \nabla\phi_\varepsilon(x).$$

Thus, the stationarity of u impies that

$$
\begin{aligned}
0 = \delta \mathcal{F}_\Lambda (u, D)[\xi] &= \int_D \left(-2\nabla u \cdot D\xi \, \nabla u + |\nabla u|^2 \operatorname{div} \xi + \Lambda \, \mathbb{1}_{\Omega_u} \operatorname{div} \xi \right) dx \\
&= \int_D \left(-2\phi_\varepsilon |\nabla u|^2 - 2(x \cdot \nabla u)(\nabla \phi_\varepsilon \cdot \nabla u) + (d\phi_\varepsilon + x \cdot \nabla \phi_\varepsilon)(|\nabla u|^2 + \Lambda \, \mathbb{1}_{\Omega_u}) \right) dx \\
&= \int_{B_r} \left((d-2)|\nabla u|^2 + d\Lambda \mathbb{1}_{\Omega_u} \right) \phi_\varepsilon \, dx + \frac{1}{\varepsilon} \int_{B_r \setminus B_{(1-\varepsilon)r}} \left(2\left(\frac{x}{|x|} \cdot \nabla u \right)^2 - |\nabla u|^2 - \Lambda \mathbb{1}_{\Omega_u} \right) dx,
\end{aligned}
$$

which passing to the limit as $\varepsilon \to 0$ implies that

$$
(d-2) \int_{B_r} |\nabla u|^2 \, dx + d\Lambda |\Omega_u \cap B_r| = r \int_{\partial B_r} \left(|\nabla_\tau u|^2 - |\nabla_\nu u|^2 + \Lambda \mathbb{1}_{\Omega_u} \right) d\mathcal{H}^{d-1}. \tag{9.9}
$$

Since $\Delta u = 0$ on Ω_u, we have that

$$
2 \int_{B_r} |\nabla u|^2 \, dx = 2 \int_{B_r} \operatorname{div}(u\nabla u) \, dx = 2 \int_{\partial B_r} u(\nu \cdot \nabla u) \, d\mathcal{H}^{d-1},
$$

which together with (9.9) implies (9.8).   □

Proposition 9.9 (Monotonicity Formula for Stationary Free Boundaries) *Suppose that D is a bounded open set in $\mathbb{R}^d$ and $u \in H^1(D)$ is a non-negative function which is stationary for $\mathcal{F}_\Lambda$ (in the sense of Definition 9.7). Let $x_0 \in D$ and $\delta_{x_0} = \operatorname{dist}(x_0, \partial D)$. Then the function $r \mapsto W_\Lambda(u_{x_0,r})$ is non-decreasing on the interval $(0, \delta_{x_0})$ and we have*

$$
\frac{\partial}{\partial r} W_\Lambda(u_{x_0,r}) \geq \frac{2}{r} \int_{\partial B_1} |x \cdot \nabla u_{x_0,r} - u_{x_0,r}|^2 \, d\mathcal{H}^{d-1}. \tag{9.10}
$$

Proof By Lemmas 9.8 and 9.2 we obtain precisely (9.10).   □

9.3   Homogeneity of the Blow-Up Limits

In this section, we use the Weiss' monotonicity formula to prove that the blow-up limits of u are one-homogeneous functions. The most general result is given in Lemma 9.10. We then prove the homogeneity of the blow-up limits of stationary functions (Lemma 9.11) and the homogeneity of the blow-up limits of minimizers of $\mathcal{F}_\Lambda$ (Proposition 9.12).

Lemma 9.10 *Suppose that D is a bounded open set in $\mathbb{R}^d$ and $u \in H^1(D)$ is a non-negative function. Let $x_0 \in D$ and $\delta_{x_0} = \operatorname{dist}(x_0, \partial D)$. Let $r_n \to 0$ be an infinitesimal sequence and let $u_n := u_{r_n, x_0} \in H^1(B_1)$. Suppose that*

(a) the limit

$$L := \lim_{r \to 0} W_\Lambda(u_{r,x_0}),$$

exists and is finite;

(b) u_n converges strongly in $H^1(B_1)$ to a function $u_\infty \in H^1(B_1)$;

(c) $\mathbb{1}_{\Omega_{u_n}}$ converges strongly in $L^1(B_1)$ to $\mathbb{1}_{\Omega_{u_\infty}}$;

(d) u_∞ is stationary for $\mathcal{F}_\Lambda$ in B_1.

Then u_∞ is one-homogeneous.

Proof Without loss of generality, we suppose that $x_0 = 0$ and we write $u_{r,x_0} = u_r$. We set for simplicity $v := u_\infty$. By the hypothesis (a), we have that,

$$L = \lim_{n \to \infty} W_\Lambda(u_{sr_n}) \qquad \text{for every} \qquad s < 0 \le 1.$$

On the other hand, the strong convergence of u_n and $\mathbb{1}_{\Omega_{u_n}}$ implies that

$$\lim_{n \to \infty} W_\Lambda(u_{sr_n}) = W_\Lambda(v_s),$$

where we recall that $v_s(x) = \dfrac{1}{s} v(sx)$. This implies that

$$W_\Lambda(v_s) = L \qquad \text{for every} \qquad s \in (0, 1],$$

and, by Proposition 9.9, we obtain that

$$0 = \frac{\partial}{\partial s} W_\Lambda(v_s) \ge \frac{2}{s} \int_{\partial B_1} |x \cdot \nabla v_s - v_s|^2 \, d\mathcal{H}^{d-1},$$

which, by Lemma 9.3, gives that v is one-homogeneous. $\qquad \square$

Lemma 9.11 (Homogeneity of the Blow-Up Limits) *Suppose that D is a bounded open set in $\mathbb{R}^d$ and $u \in H^1(D)$ is a non-negative function which is stationary for $\mathcal{F}_\Lambda$ (in the sense of Definition 9.7). Let $x_0 \in D \cap \partial \Omega_u$, $r_n \to 0$ be an infinitesimal sequence and $u_n := u_{r_n, x_0} \in H^1(B_1)$. Suppose that*

(a) u_n converges strongly in $H^1(B_1)$ to a function $u_\infty \in H^1(B_1)$;

(b) $\mathbb{1}_{\Omega_{u_n}}$ converges strongly in $L^1(B_1)$ to $\mathbb{1}_{\Omega_{u_\infty}}$.

Then u_∞ is one-homogeneous.

Proof Since u is stationary, Lemma 9.9 implies that the function $r \mapsto W_\Lambda(u_{x_0,r})$ is non-decreasing in r. Thus, the limit

$$L := \lim_{r \to 0} W_\Lambda(u_{x_0,r}) = \inf_{r>0} W_\Lambda(u_{x_0,r}),$$

does exist and so the hypothesis (a) of Lemma 9.10 is fulfilled. Now, the strong convergence of u_n and $\mathbb{1}_{\Omega_{u_n}}$ to u_∞ and $\mathbb{1}_{\Omega_{u_\infty}}$ in B_1, and the definition of the first variation $\delta\mathcal{F}_\Lambda(\cdot, D)$ imply that u_∞ is also stationary in B_1. Thus, hypothesis (d) of Lemma 9.10 is also fulfilled and, so the claim follows by Lemma 9.10. $\qquad\square$

Proposition 9.12 (Homogeneity of the Blow-Up Limits) *Suppose that D is a bounded open set in $\mathbb{R}^d$ and $u \in H^1(D)$ is a non-negative function and a local minimizer of $\mathcal{F}_\Lambda$ in D. Let $x_0 \in D$. Then every blow-up limit $u_0 \in \mathcal{BU}_u(x_0)$ is one-homogeneous.*

Proof By Lemma 9.6, every minimizer of $\mathcal{F}_\Lambda$ is stationary for $\mathcal{F}_\Lambda$. Moreover, by Proposition 6.2, we have that the conditions (a) and (b) of Lemma 9.11 are fulfilled. This concludes the proof. $\qquad\square$

9.4   Regularity of the Free Boundaries in Dimension Two

The main result of this section is the following.

Proposition 9.13 (One-Homogeneous Global Minimizers in Dimension Two) *Let $z : \mathbb{R}^2 \to \mathbb{R}$ be a one-homogeneous global minimizer of $\mathcal{F}_\Lambda$ in $\mathbb{R}^2$. Then, there is $v \in \mathbb{R}^2$ such that*

$$z(x) = \sqrt{\Lambda}\,(x \cdot v)_+ \quad \text{for every} \quad x \in \mathbb{R}^2.$$

In particular, we obtain that the critical dimension d^* is at least 3 (see Definition 1.5).
The proof of Proposition 9.13 is based on the following lemma.

Lemma 9.14 *Let $z \in H^1_{loc}(\mathbb{R}^d)$ be a continuous and non-negative one-homogeneous function in $\mathbb{R}^d$. Then,*

$$\Delta z = 0 \quad \text{in} \quad \Omega_z,$$

if and only if, the trace $c = z|_{\partial B_1} \in H^1(\partial B_1)$ is such that

$$-\Delta_{\mathbb{S}}c = (d-1)c \quad \text{in the (open) set} \quad \Omega_c \cap \partial B_1.$$

Proof The proof follows simply by writing the Laplacian in polar coordinates. In fact, we have that $z(r, \theta) = rc(\theta)$ and

$$\Delta z(r, \theta) = \partial_{rr} z(r, \theta) + \frac{d-1}{r} \partial_r z(r, \theta) + \frac{1}{r^2} \Delta_{\mathbb{S}} z(r, \theta)$$
$$= \frac{1}{r} \big((d-1) c(\theta) + \Delta_{\mathbb{S}} c(\theta) \big),$$

which concludes the proof of Lemma 9.14.   □

Proof of Proposition 9.13 Let $z(r, \theta) = rc(\theta)$ and let $\Omega_c \subset \mathbb{S}^1$ be the set $\{c > 0\}$. Since c is continuous (see Sect. 3), we have that Ω_c is open and so it is a countable union of disjoint arcs (which we identify with segments on the real line). Notice that $\Omega_c \neq \mathbb{S}^1$ since $z(0) = 0$ and z minimizes locally $\mathcal{F}_\Lambda$ (the local minimizers cannot have isolated zeros, for instance, by the density estimates from Sect. 5.1). Now, Lemma 9.14 implies that on each arc $\mathcal{I} \subset \Omega_c$, the trace c is a solution of the PDE

$$-c''(\theta) = c(\theta) \quad \text{in} \quad \mathcal{I}, \qquad c > 0 \quad \text{in} \quad I, \qquad c = 0 \quad \text{on} \quad \partial I.$$

Thus, up to a translation $\mathcal{I} = (0, \pi)$ and $c(\theta)$ is a multiple of $\sin \theta$ on $\mathcal{I}$. Thus, Ω_c is a union of disjoint arcs, each one of length π. Thus, these arcs can be at most two. Now, by Lemma 2.9 and the fact that $0 \in \partial \Omega_z$, we get that $|\Omega_z \cap B_1| < |B_1| = \pi$ and so, $\mathcal{H}^1(\Omega_c) < 2\pi$. This means that Ω_c is an arc of length π and that z is of the form $z(x) = a (x \cdot v)$, for some constant $a > 0$. Since z is a local minimizer in $\mathbb{R}^d$ and $\partial \Omega_z$ is smooth, Lemma 6.11 implies that $a = \sqrt{\Lambda}$, which concludes the proof.   □

9.5 The Optimality Condition on the Free Boundary: A Monotonicity Formula Approach

The aim of this subsection is to give an alternative proof to the fact that the (local) minimizers of $\mathcal{F}_\Lambda$ are viscosity solutions to the problem

$$\Delta u = 0 \quad \text{in} \quad \Omega_u, \qquad |\nabla u| = \sqrt{\Lambda} \quad \text{on} \quad \partial \Omega_u.$$

The main result of the subsection is Proposition 9.18, which can be applied not only to minimizers, but also to measure constrained minimizers (see Theorem 1.9 and Sect. 11). It can also be applied to a large class of problems in which a monotonicity formula does hold. In fact, the proof is quite robust and can be applied to almost-minimizers (see for instance [46]) and to vectorial problems (see [41]), for which the construction of competitors is typically more involved. The proof of Proposition 9.18 is based on the following two lemmas. Before we give the two statements, we recall that, for any $d \geq 2$, we identify the $(d-1)$-dimensional sphere $\mathbb{S}^{d-1}$ with the boundary of the unit ball ∂B_1 in $\mathbb{R}^d$. In particular, we will use the notation

$$\mathbb{S}^{d-1}_+ = \{x := (x_1, \ldots, x_d) \in \partial B_1 \subset \mathbb{R}^d \ : \ x_d > 0\}.$$

Lemma 9.15 *Suppose that $c \in H^1(\mathbb{S}^{d-1})$ is a continuous non-negative and non-constantly-vanishing function, satisfying the following conditions:*

(a) $\Omega_c \subset \mathbb{S}^{d-1}_+$, where as usual $\Omega_c := \{c > 0\}$;
(b) $\Delta_{\mathbb{S}} c + (d-1)c = 0$ in Ω_c.

Then, $\Omega_c = \mathbb{S}^{d-1}_+$ and there is a constant $\alpha > 0$ such that

$$c(x) = \alpha(x \cdot e_d)_+ \qquad \text{for every} \qquad x \in \partial B_1.$$

Lemma 9.16 *Suppose that $c \in H^1(\mathbb{S}^{d-1})$ is a continuous non-negative function, satisfying the following conditions:*

(a) $\mathbb{S}^{d-1}_+ \subset \Omega_c = \{c > 0\}$;
(b) $\Delta_{\mathbb{S}} c + (d-1)c = 0$ in Ω_c.

Then, c is given by one of the following functions:

(i) $c(x) = \alpha(x \cdot e_d)_+$, where $\alpha > 0$ is a positive constant;
(ii) $c(x) = \alpha(x \cdot e_d)_+ + \beta(x \cdot e_d)_-$, where $\alpha > 0$ and $\beta > 0$.

In the proofs of Lemmas 9.15 and 9.16 we will use the following well-known result, whose proof we the leave to the reader.

Lemma 9.17 (Variational Characterization of the Principal Eigenvalue) *Let $\Omega \subset \mathbb{S}^{d-1}$ be a connected open subset of the unit sphere. Let $\phi \in H^1_0(\Omega)$ be a given non-zero function. Then, the following are equivalent:*

(i) $\phi > 0$ in Ω, $\int_\Omega \phi^2 d\theta = 1$, and there is $\lambda \geq 0$ for which ϕ solves the PDE

$$-\Delta_{\mathbb{S}}\phi = \lambda\phi \quad in \quad \Omega$$

in the usual weak sense:

$$\int_\Omega \nabla_\theta \phi \cdot \nabla_\theta \eta \, d\theta = \lambda \int_\Omega \phi\eta \, d\theta \quad for \ every \quad \eta \in H^1_0(\Omega);$$

(ii) ϕ is the unique (up to a sign) solution of the variational problem

$$\min\left\{ \int_\Omega |\nabla_\theta \psi|^2 \, d\theta \ : \ \psi \in H^1_0(\Omega), \ \int_\Omega \psi^2 \, d\theta = 1 \right\}.$$

Proof of Lemma 9.15 Since the linear functions are one-homogeneous and harmonic in $\mathbb{R}^d$, we have that the function

$$\phi_1(\theta) = (\theta \cdot e_d)_+,$$

defined on the sphere solves the equation

$$-\Delta_{\mathbb{S}}\phi_1 = (d - 1)\phi_1 \quad \text{in} \quad \mathbb{S}^{d-1}_+.$$

In particular, setting $\alpha_d := \left(\int_{\mathbb{S}^{d-1}} \phi_1^2 \, d\theta \right)^{-1}$, we get that $\alpha_d\phi_1$ is the unique minimizer of

$$d - 1 = \min \left\{ \int_{\mathbb{S}^{d-1}_+} |\nabla_\theta \psi|^2 \, d\theta \; : \; \psi \in H^1_0(\mathbb{S}^{d-1}_+), \int_{\mathbb{S}^{d-1}_+} \psi^2 \, d\theta = 1 \right\}.$$

On the other hand, $c \in H^1_0(\mathbb{S}^{d-1}_+)$ and solves the equation $-\Delta_{\mathbb{S}}c = (d - 1)c$ in Ω_c. Thus,

$$\int_{\mathbb{S}^{d-1}_+} |\nabla_\theta c|^2 \, d\theta = \int_{\Omega_c} |\nabla_\theta c|^2 \, d\theta = (d - 1) \int_{\Omega_c} c^2 \, d\theta = (d - 1) \int_{\mathbb{S}^{d-1}_+} c^2 \, d\theta,$$

which means that (up to a multiplicative constant) c is a solution of the same problem. Thus, the uniqueness of ϕ_1 gives the claim. $\square$

Proof of Lemma 9.16 Let $\widetilde{\Omega}_c$ be the connected component of Ω_c containing $\mathbb{S}^{d-1}_+$; and let $\widetilde{c}$ be the restriction of c to $\widetilde{\Omega}_c$. Thus, $\widetilde{\Omega}_c = \{\widetilde{c} > 0\}$ and $\widetilde{c}$ solves the PDE

$$-\Delta_{\mathbb{S}}\widetilde{c} = (d - 1)\widetilde{c} \quad \text{in} \quad \widetilde{\Omega}_c.$$

Thus, $\widetilde{c}$ is the unique minimizer of

$$d - 1 = \min \left\{ \int_{\widetilde{\Omega}_c} |\nabla_\theta \psi|^2 \, d\theta \; : \; \psi \in H^1_0(\widetilde{\Omega}_c), \int_{\widetilde{\Omega}_c} \psi^2 \, d\theta = 1 \right\}.$$

Thus, reasoning as in the proof of Lemma 9.15, we get that $\widetilde{\Omega}_c = \mathbb{S}^{d-1}_+$ and that there is a constant $\alpha > 0$ such that

$$\widetilde{c}(\theta) = \alpha(\theta \cdot e_d)_+.$$

We now consider two cases. If Ω_c has only one connected component, then $\Omega_c = \widetilde{\Omega}_c$ and $c = \widetilde{c}$, which concludes the proof. If Ω_c has more than one connected components, then $\Omega_c \setminus \widetilde{\Omega}_c$ is non-empty and is contained in the half-sphere

$$\mathbb{S}^{d-1}_- = \{x := (x_1, \ldots, x_d) \in \partial B_1 \subset \mathbb{R}^d \; : \; x_d < 0\}.$$

Thus, applying Lemma 9.15, we get that the restriction of c on $\Omega_c \setminus \widetilde{\Omega}_c$ should be of the form $\beta(\theta \cdot e_d)_-$, for some positive constant β, which concludes the proof. $\square$

Proposition 9.18 *Suppose that $D \subset \mathbb{R}^d$ is a bounded open set and that $u \in H^1(D)$ is a continuous non-negative function such that:*

(a) u is harmonic in $\Omega_u = \{u > 0\}$.
(b) Ω_u satisfies the upper density bound

$$\limsup_{r \to 0} \frac{|\Omega_u \cap B_r(x_0)|}{|B_r|} < 1 \qquad for\ every \qquad x_0 \in \partial\Omega_u \cap D.$$

(c) For every $x_0 \in D \cap \partial\Omega_u$ and every infinitesimal sequence $r_n \to 0$, there is a subsequence (that we still denote by r_n) such that the blow-up sequence u_{r_n, x_0} converges uniformly in B_1 to a blow-up limit $u_0 : B_1 \to \mathbb{R}$ ($u_0 \in \mathcal{BU}_u(x_0)$).
(d) Every blow-up limit $\mathcal{BU}_u(x_0) \ni u_0 : B_1 \to \mathbb{R}$ is a one-homogeneous non-identically-zero function, which is stationary for the functional $\mathcal{F}_\Lambda$.

Then u satisfies the optimality condition

$$|\nabla u| = \sqrt{\Lambda} \quad on \quad \partial\Omega_u \cap D,$$

in viscosity sense.

Proof Suppose first that the function φ touches u from below in $x_0 \in \partial\Omega_u$ and assume that $x_0 = 0$. Consider the blow-up sequences

$$u_n(x) = \frac{1}{r_n} u(r_n x) \qquad and \qquad \varphi_n(x) = \frac{1}{r_n} \varphi(r_n x),$$

as $r_n \to 0$, the condition (c) implies that, up to a subsequence, we have

$$u_0 = \lim_{n \to \infty} u_n(x) \qquad and \qquad \varphi_0 = \lim_{n \to \infty} \varphi_n(x), \tag{9.11}$$

the convergence being uniform in B_1. In particular, since u_n are harmonic in Ω_{u_n}, the uniform convergence of u_n to u_0 implies that also u_0 is harmonic on Ω_{u_0}.

Notice that, as φ is smooth, we have $\varphi_0(x) = \xi \cdot x$, where the vector $\xi \in \mathbb{R}^d$ is precisely the gradient $\nabla\varphi(0)$. Without loss of generality we may assume that $\xi = A e_d$ for some constant $A \geq 0$, thus

$$|\nabla\varphi(0)| = |\nabla\varphi_0(0)| = A \qquad and \qquad \varphi_0(x) = A x_d. \tag{9.12}$$

Moreover, we can assume that $A > 0$ since otherwise the inequality $|\nabla\varphi| \leq \sqrt{\Lambda}$ holds trivially.

Now, since $u_0 \geq \varphi_0$, we obtain that $u_0 > 0$ on the set $\{x_d > 0\}$. Thus, u_0 is a 1-homogeneous harmonic function on the cone $\{u_0 > 0\} \supset \{x_d > 0\}$. By

Lemma 9.16, there are only two possibilities:

$$u_0(x) = \alpha x_d^+ \qquad \text{or} \qquad u_0(x) = \alpha x_d^+ + \beta x_d^- \,.$$

The second case is ruled out since it contradicts (b). Thus,

$$u_0(x) = \alpha x_d^+ \quad \text{for every} \quad x \in B_1. \tag{9.13}$$

Now, the stationarity of u_0 (condition (d)) and Lemma 9.5 imply that $\alpha = \sqrt{\Lambda}$. By the inequality $u_0 \geq \varphi_0$, we get that $\sqrt{\Lambda} \geq A$.

   Suppose now that φ touches u from above at a point x_0 and assume that $x_0 = 0$. Again, we consider the blow-up limits U_0 and φ_0 defined in (9.11) and we assume that φ_0 is given by (9.12). Since u_0 is not identically zero (assumption (d)), we get that $a > 0$. Since $u_0 \leq \varphi_0$ we have that the set $\{u_0 > 0\}$ is contained in the half-space $\{x_d > 0\}$. By the one-homogeneity of u_0 and Lemma 9.15 we obtain that necessarily $\{u_0 > 0\} = \{x_d > 0\}$. Thus, u_0 is of the form (9.13) for some $\alpha > 0$. Now, the stationarity of u_0 implies that necessarily $\alpha = \sqrt{\Lambda}$ and, since $u_0 \leq \varphi_0$, we get that $|\nabla\varphi(0)| = A \geq \sqrt{\Lambda}$, which concludes the proof. $\square$

9.6   Energy and Lebesgue Densities

In this section, we prove that if u is a (local) minimizer of $\mathcal{F}_\Lambda$, then at every boundary point $x_0 \in \partial\Omega_u$ the Lebesgue density of the set Ω_u is well-defined. Moreover, we characterize the regular part of the free boundary in terms of the Lebesgue density. Most of the ideas in this section come from [41], where we used a similar characterization of the regular part of the vectorial free boundaries. In the case of the one-phase problem, we will not use this result in the proofs of neither of the Theorems 1.2, 1.4, 1.9 nor 1.10; we give it here only for the sake of completeness. The precise statement is the following:

Proposition 9.19 *Suppose that $D \subset \mathbb{R}^d$ is a bounded open set and that $u \in H^1(D)$ is a non-negative function, a local minimizer of $\mathcal{F}_\Lambda$ in D. Then, the limit*

$$\lim_{r \to 0} \frac{|\Omega_u \cap B_r(x_0)|}{|B_r|} \quad \text{exists, for every} \quad x_0 \in \partial\Omega_u \cap D. \tag{9.14}$$

Thus, we can write

$$\partial\Omega_u \cap D = \bigcup_{\frac{1}{2} \leq \gamma < 1} \Omega_u^{(\gamma)} \cap D. \tag{9.15}$$

The regular and the singular parts of the free boundary are given by

$$\mathrm{Reg}\,(\partial\Omega_u) \cap D = \Omega_u^{(1/2)} \cap D \qquad and \qquad \mathrm{Sing}\,(\partial\Omega_u) \cap D = \bigcup_{\frac{1}{2} < \gamma < 1} \Omega_u^{(\gamma)} \cap D.$$

$$(9.16)$$

Moreover, for every $\gamma \in [1/2, 1)$, we have

$$\Omega_u^{(\gamma)} \cap D = \left\{ x \in \partial\Omega_u \cap D \;:\; |\Omega_{u_0} \cap B_1| = \omega_d \gamma, \;\; for\,every \;\; u_0 \in \mathcal{BU}_u(x) \right\}. \qquad (9.17)$$

Proof The claims (9.14), (9.15) and (9.17) follow directly by Lemma 9.20 below. The claim (9.16), follows by Lemma 9.22. $\qquad\qquad\square$

Lemma 9.20 (Energy and Lebesgue Densities) *Suppose that $D \subset \mathbb{R}^d$ is a bounded open set and that $u \in H^1(D)$ is a continuous non-negative function such that:*

(a) *For every $x_0 \in D$ and every infinitesimal sequence $r_n \to 0$, there is a subsequence (that we still denote by r_n) such that:*

- $u_n := u_{r_n, x_0}$ *converges strongly in $H^1(B_1)$ to a function $u_0 : B_1 \to \mathbb{R}$;*
- $\mathbb{1}_{\Omega_{u_n}}$ *converges in $L^2(B_1)$ to $\mathbb{1}_{\Omega_{u_0}}$.*

(As usual, we say that u_0 is a blow-up limit of u, and we note $u_0 \in \mathcal{BU}_u(x_0)$.)

(b) *Every blow-up limit $\mathcal{BU}_u(x_0) \ni u_0 : B_1 \to \mathbb{R}$ is a one-homogeneous non-identically-zero function such that $\Delta u_0 = 0$ in $\Omega_{u_0} \cap B_1$.*

(c) *For every $x_0 \in \partial\Omega_u \cap D$, the limit*

$$\Theta(u, x_0) := \lim_{r \to 0} W_\Lambda(u_{r, x_0}),$$

does exist.

Then, for every $x_0 \in \partial\Omega_u \cap D$, we have that

$$\frac{1}{\Lambda \omega_d} \Theta(u, x_0) = \lim_{r \to 0} \frac{|\Omega_u \cap B_r(x_0)|}{|B_r|}.$$

Moreover, for every $u_0 \in \mathcal{BU}_u(x_0)$, we have that

$$\frac{1}{\Lambda \omega_d} \Theta(u, x_0) = \frac{|\Omega_{u_0} \cap B_1|}{|B_1|} = \frac{1}{\Lambda \omega_d} W_\Lambda(u_0).$$

Proof We first notice that (b) implies that

$$W_\Lambda(u_0) = \Lambda |\Omega_{u_0} \cap B_1|.$$

Let $x_0 \in \partial\Omega_u \cap D$ and the infinitesimal sequence $r_n \to 0$ be given. Then, by (a), up to a subsequence, u_{r_n,x_0} converges to a blow-up limit u_0. Using (c) and then again (a), we get

$$\lim_{r \to 0} W_\Lambda(u_{r,x_0}) = \lim_{n \to \infty} W_\Lambda(u_{r_n,x_0}) = W_\Lambda(u_0).$$

On the other hand, the strong $H^1(B_1)$ convergence of u_{r_n,x_0} to u_0 implies that

$$\lim_{n \to \infty} W_0(u_{r_n,x_0}) = W_0(u_0) = 0.$$

Then, we have

$$|\Omega_{u_0} \cap B_1| = \frac{1}{\Lambda} \lim_{n \to \infty} W_\Lambda(u_{r_n,x_0}) = \lim_{n \to \infty} |\{u_{r_n,x_0} > 0\} \cap B_1| = \lim_{n \to \infty} \frac{|\Omega_u \cap B_{r_n}(x_0)|}{r_n^d}$$

which concludes the proof. □

In the proof of Lemma 9.22, we will use the following result.

Theorem 9.21 (The Spherical Caps Minimize λ_1 on the Sphere) *For any (quasi-)open spherical set $\Omega \subset \mathbb{S}^{d-1}$ we define the first eigenvalue $\lambda_1(\Omega)$ as*

$$\lambda_1(\Omega) := \inf\left\{ \int_\Omega |\nabla_\theta c|^2 \, d\theta \ : \ \int_\Omega c^2(\theta) \, d\theta = 1, \ c \in H_0^1(\Omega) \right\}.$$

For every open set $\Omega \subset \mathbb{S}^{d-1}$ such that $\mathcal{H}^{d-1}(\Omega) \le \frac{1}{2} d\omega_d$ we have that

$$\lambda_1(\Omega) \ge \lambda_1(\mathbb{S}_+^{d-1}),$$

with equality if and only if, up to a rotation, $\Omega = \mathbb{S}_+^{d-1}$.

Lemma 9.22 (Characterization of the Regular Part of the Free Boundary) *Suppose that $D \subset \mathbb{R}^d$ is a bounded open set and that $u \in H^1(D)$ is as in Lemma 9.20. Then,*

$$\lim_{r \to 0} \frac{|\Omega_u \cap B_r(x_0)|}{|B_r|} \ge \frac{1}{2} \qquad \text{for every} \qquad x_0 \in \partial\Omega_u \cap D. \tag{9.18}$$

Moreover,

$$\lim_{r \to 0} \frac{|\Omega_u \cap B_r(x_0)|}{|B_r|} = \frac{1}{2},$$

if and only if, every blow-up limit $u_0 \in \mathcal{BU}_u(x_0)$ is of the form

$$u_0(x) = (v \cdot x)_+ \qquad \text{for some} \qquad v \in \mathbb{R}^d. \tag{9.19}$$

In particular, if u is a minimizer of $\mathcal{F}_\Lambda$ in D, then $Reg(\partial\Omega_u) = \Omega_u^{(1/2)}$ in D.

Proof Suppose that $x_0 \in \partial\Omega_u \cap D$ and let

$$\gamma := \lim_{r \to 0} \frac{|\Omega_u \cap B_r(x_0)|}{|B_r|}.$$

Let $r_n \to 0$ be an infinitesimal sequence. Then, by the assumption Lemma 5.1 (a), up to extracting a subsequence, we can suppose that u_{r_n,x_0} converges to a blow-up limit $u_0 : \mathbb{R}^d \to \mathbb{R}$. By the hypothesis Lemma 5.1 (b), we get that u_0 is one-homogeneous and harmonic in $\Omega_{u_0} \cap B_1$. This implies that, on the sphere ∂B_1, u_0 solves the PDE

$$\Delta_S u_0 = (d-1)u_0 \quad \text{in} \quad \Omega_{u_0} \cap \partial B_1.$$

Thus, Theorem 9.21 implies that

$$\mathcal{H}^{d-1}(\Omega_u \cap \partial B_1) \geq \frac{d\omega_d}{2},$$

which by the homogeneity of u_0 gives that

$$|\Omega_{u_0} \cap B_1| \geq \frac{\omega_d}{2}.$$

Now, the convergence of $\Omega_{u_{r_n,x_0}}$ to Ω_{u_0} implies that

$$\gamma = \lim_{n \to \infty} \frac{|\Omega_u \cap B_{r_n}(x_0)|}{|B_{r_n}|} = \lim_{n \to \infty} \frac{|\Omega_{u_{r_n,x_0}} \cap B_1|}{|B_1|} = \frac{|\Omega_{u_0} \cap B_1|}{|B_1|} \geq \frac{1}{2},$$

which concludes the proof of the lower bound (9.18). In the case of equality $\gamma = 1/2$, we have that $u_0\big|_{\partial B_1}$ is precisely the first eigenvalue on the half-sphere $\mathbb{S}_{d-1}^+$, whose one-homogeneous extension is precisely (9.19). $\qquad\square$

Chapter 10
Dimension of the Singular Set

In this chapter, we prove Theorem 1.4. As in the original work of Weiss (see [52]), we will use the so-called Federer's dimension reduction principle, which first appeared in [32].

This chapter is organized as follows.

- In Sect. 10.1 we give the definitions of the Hausdorff measure and Hausdorff dimension; we also state and prove the main properties of the Hausdorff measure, which we will need for the proof of Theorem 1.4.
- In Sect. 10.2 we give a general result for the convergence of the singular sets of a sequence of functions.
- In Sect. 10.3 we study the structure of the singular set of the one-homogeneous global minimizers of $\mathcal{F}_\Lambda$.
- Finally, in Sect. 10.4, we use the results of the previous subsections (Lemmas 10.7 and 10.12) to prove Theorem 1.4.

10.1 Hausdorff Measure and Hausdorff Dimension

In this section we define the notions of Hausdorff measure and Hausdorff dimension and we also give their main properties. For more details, we refer to the book [31].

We recall that, for every $s > 0$, $\delta \in (0, +\infty]$ and every set $E \subset \mathbb{R}^d$,

$$\mathcal{H}^s_\delta(E) := \frac{\omega_s}{2^s} \inf \left\{ \sum_{j=1}^\infty \left(\operatorname{diam} U_j \right)^s \; : \; \text{for every family of sets } \{U_j\}_{j=1}^\infty \right.$$

$$\left. \text{such that } E \subset \bigcup_{j=1}^\infty U_j \text{ and } \operatorname{diam} U_j \leq \delta, \text{ for every } j \geq 1 \right\},$$

$$(10.1)$$

© The Author(s) 2023
B. Velichkov, *Regularity of the One-phase Free Boundaries*,
Lecture Notes of the Unione Matematica Italiana 28,
https://doi.org/10.1007/978-3-031-13238-4_10

where, for any $s \in (0, +\infty)$, the constant ω_s is defined as

$$\omega_s := \frac{\pi^{s/2}}{\Gamma(s/2 + 1)} \quad \text{where} \quad \Gamma(s) := \int_0^{+\infty} x^{s-1} e^x \, dx.$$

Definition 10.1 (Hausdorff Measure) For any $s \geq 0$, $\mathcal{H}^s(E)$ denotes the s-dimensional Hausdorff measure of a set $E \subset \mathbb{R}^d$ and is defined as:

$$\mathcal{H}^s(E) := \lim_{\delta \to 0_+} \mathcal{H}^s_\delta(E) = \sup_{\delta > 0} \mathcal{H}^s_\delta(E).$$

Remark 10.2 The constant in (10.1) is chosen in such a way that we have

$$\mathcal{H}^d(B_r) = |B_r| = \omega_d r^d \quad \text{and} \quad \mathcal{H}^{d-1}(\partial B_r) = d\omega_d r^{d-1}.$$

Definition 10.3 The Hausdorff dimension of a set $E \subset \mathbb{R}^d$ is defined as

$$\dim_{\mathcal{H}} E := \inf \{ s > 0 \; : \; \mathcal{H}^s(E) = 0 \}.$$

The following elementary properties of the Hausdorff measure are an immediate consequence of the definitions of $\mathcal{H}^s$, $\mathcal{H}^s_\delta$ and $\mathcal{H}^s_\infty$.

Proposition 10.4 (Properties of the Hausdorff Measure)

(i) *For every $s > 0$ and $\delta \in (0, \infty]$, the set functionals $\mathcal{H}^s$ and $\mathcal{H}^s_\delta$ are translation invariant and increasing with respect to the set inclusion. Moreover, we have*

$$\mathcal{H}^s(rE) = r^s \mathcal{H}^s(E) \quad \text{and} \quad \mathcal{H}^s_\infty(rE) = r^s \mathcal{H}^s_\infty(E) \quad \text{for any} \quad E \subset \mathbb{R}^d \quad \text{and} \quad r > 0.$$

(ii) *The function $\delta \mapsto \mathcal{H}^s_\delta$ is non-decreasing in δ. In particular, we have*

$$\mathcal{H}^s(E) \leq \mathcal{H}^s_\delta(E) \leq \mathcal{H}^s_\infty(E) \quad \text{for any} \quad E \subset \mathbb{R}^d \quad \text{and any} \quad \delta > 0.$$

(iii) *Given $s > 0$ and $E \subset \mathbb{R}^d$, we have that*

$$\mathcal{H}^s(E) = 0 \quad \text{if and only if} \quad \mathcal{H}^s_\infty(E) = 0.$$

(iv) *Given a sequence of sets $E_j \subset \mathbb{R}^d$, $s > 0$ and $\delta \in (0, +\infty]$ we have that*

$$\mathcal{H}^s_\delta(E) \leq \sum_{j=1}^{\infty} \mathcal{H}^s_\delta(E_j) \quad \text{where} \quad E = \bigcup_{j=1}^{\infty} E_j.$$

In particular, $\mathcal{H}^s(E) = 0$ if and only if $\mathcal{H}^s(E_j) = 0$, for every $j \geq 1$.

Lemma 10.5 (Existence of Points of Positive Density) *Let $s > 0$ and let $K \subset \mathbb{R}^d$ be a given set. If $\mathcal{H}^s(K) > 0$, then there is a point $x_0 \in K$ such that*

$$\limsup_{r \to 0} \frac{\mathcal{H}^s(K \cap B_r(x_0))}{r^s} > 0. \tag{10.2}$$

Proof Suppose that (10.2) does not hold. Then, we have

$$\limsup_{r \to 0} \frac{\mathcal{H}^s(K \cap B_r(x_0))}{r^s} = 0. \tag{10.3}$$

Let $K_{\delta,\varepsilon} \subset K$ be the set

$$K_{\delta,\varepsilon} = \left\{ x \in K \ : \ \mathcal{H}^s(K \cap B_r(x)) \leq \varepsilon r^s \quad \text{for every} \quad r \leq \delta \right\}.$$

By (10.3), we have that

$$\bigcup_{\delta>0} K_{\delta,\varepsilon} = \bigcup_{n=1}^{\infty} K_{\delta,1/n} = K \qquad \text{for every fixed} \qquad \varepsilon > 0. \tag{10.4}$$

Let now δ and ε be fixed and let $\{U_i\}_{i \geq 1}$ be a family of sets of diameter $\operatorname{diam} U_i \leq \delta$ such that $K_{\delta,\varepsilon} \subset \bigcup_i U_i$. Then, the subadditivity of $\mathcal{H}^s_\delta$ gives that

$$\mathcal{H}^s_\delta(K_{\delta,\varepsilon}) \leq \sum_{i=1}^{\infty} \mathcal{H}^s_\delta(U_i \cap K_{\delta,\varepsilon}) \leq \sum_{i=1}^{\infty} \mathcal{H}^s(U_i \cap K_{\delta,\varepsilon})$$

$$\leq \sum_{i=1}^{\infty} \mathcal{H}^s(U_i \cap K) \leq \sum_{i=1}^{\infty} \varepsilon (\operatorname{diam} U_i)^s,$$

where the last inequality holds since the set $U_i \cap K$ is contained in a ball of radius

$$r_i = \operatorname{diam} U_i \leq \delta.$$

Taking the infimum over all coverings C_i with sets of diameter less than or equal to δ, we get that

$$\mathcal{H}^s_\delta(K_{\delta,\varepsilon}) \leq \varepsilon \frac{2^s}{\omega_s} \mathcal{H}^s_\delta(K_{\delta,\varepsilon}),$$

and so, for ε small enough, $\mathcal{H}^s_\delta(K_{\delta,\varepsilon}) = 0$, which implies that $\mathcal{H}^s(K_{\delta,\varepsilon}) = 0$. Finally, (10.4) and the subadditivity of $\mathcal{H}^s$ imply that $\mathcal{H}^s(K) = 0$, which is a contradiction. $\qquad\square$

Lemma 10.6 (Dimension Reduction: Lemma I) *Let $s > 0$. Let $E \subset \mathbb{R}^{d-1}$ be a given set and let $\tilde{E} = E \times \mathbb{R} \subset \mathbb{R}^d$. If $\mathcal{H}^s(E) = 0$, then also $\mathcal{H}^{s+1}(\tilde{E}) = 0$.*

Proof We will prove that $\mathcal{H}^{s+1}(E \times [0, T]) = 0$ for every $T > 0$. In fact, this implies that $\mathcal{H}^{s+1}(E \times [-T, T]) = 0$ and since

$$\tilde{E} = \bigcup_{T>0} E \times [-T, T],$$

we get $\mathcal{H}^{s+1}(\tilde{E}) = 0$.

Since $\mathcal{H}^s(E) = 0$, for every $\varepsilon > 0$, there is a family of balls $B'_{r_i}(x_i) \subset \mathbb{R}^{d-1}$ such that

$$E \subset \bigcup_{i \geq 1} B'_{r_i}(x_i) \qquad \text{and} \qquad \sum_{i=1}^{\infty} r_i^s \leq \varepsilon.$$

Let now T be fixed. For every $i \in \mathbb{N}$, we consider the point $x_{i,k} \in \mathbb{R}^d$ of coordinates $x_{i,k} = (x_i, kr_i)$, for $k = 0, 1, \ldots, K_i$, where $K_i := [T/r_i] + 1$ and the family of balls $B_{2r_i}(x_{i,k})$. Notice that

$$x' \times [0, T] \subset \bigcup_k B_{2r_i}(x_{i,k}) \qquad \text{for every} \qquad x' \in B'_{r_i}(x_i) \subset \mathbb{R}^{d-1}.$$

Thus, the family of balls $\{B_{2r_i}(x_{i,k})\}_{i,k}$ is a covering of $E \times [0, T]$. We now estimate,

$$\mathcal{H}^{s+1}_{\infty}(E \times [0, T]) \leq \sum_{i=1}^{\infty} \sum_{k=1}^{K_i} (2r_i)^{s+1} = 2^{s+1} \sum_{i=1}^{\infty} \sum_{k=1}^{K_i} r_i^{s+1}$$

$$= 2^{s+1} \sum_{i=1}^{\infty} (K_i + 1) r_i^{s+1} \leq 2^{s+1} \sum_{i=1}^{\infty} \frac{2T}{r_i} r_i^{s+1},$$

where the last inequality follows by the fact that, for T large enough,

$$K_i + 1 \leq \frac{T}{r_i} + 2 \leq \frac{2T}{r_i}.$$

Thus, we get

$$\mathcal{H}^{s+1}_{\infty}(E \times [0, T]) \leq 2^{s+2} T \sum_{i=1}^{\infty} r_i^s \leq 2^{s+2} T \varepsilon,$$

which concludes the proof.   □

10.2   Convergence of the Singular Sets

In this section we will prove a general result (Lemma 10.7) for the convergence of the singular sets, which applies both to minimizers of $\mathcal{F}_\Lambda$ (Theorem 1.4) and to measure-constrained minimizers (Theorem 1.9). Recall that, if $D \subset \mathbb{R}^d$ is an open set, $u : D \to \mathbb{R}$ a given (continuous and non-negative) function, then for every ball $B_r(x_0) \subset D$, we define

$$u_{x_0,r} : B_1 \to \mathbb{R}, \qquad u_{x_0,r}(x) = \frac{1}{r} u(x_0 + rx).$$

We say that a boundary point $x_0 \in \partial\Omega_u \cap D$ is regular (and we write $x_0 \in Reg(\partial\Omega_u)$), if there is a sequence $r_n \to 0$ such that

$$\lim_{n\to\infty} \|u_{x_0,r_n} - h_v\|_{L^\infty(B_1)} = 0,$$

where for simplicity we set

$$h_v(x) = \sqrt{\Lambda} \, (x \cdot v)_+ ,$$

and we recall that

$$\left\| u_{x_0,r_n} - h_v \right\|_{L^\infty(B_1)} = \|u(x) - h_v(x - x_0)\|_{L^\infty_x(B_r(x_0))}.$$

We say that a point x_0 is singular if it is not regular, that is, if

$$x_0 \in Sing(\partial\Omega_u) := (\partial\Omega_u \cap D) \setminus Reg(\partial\Omega_u).$$

Lemma 10.7 (Convergence of the Singular Sets) *Suppose that $D \subset \mathbb{R}^d$ is a bounded open set. Let $u_n : D \to \mathbb{R}$ be a sequence of continuous non-negative functions satisfying the following conditions:*

(a) **Uniform ε-regularity.** *There are constants $\varepsilon > 0$ and $R > 0$ such that the following holds:*
   if $n \in \mathbb{N}$, $x_0 \in \partial\Omega_{u_n} \cap D$ and $r \in (0, R)$ are such that $B_r(x_0) \subset D$ and

$$\|u_n - h_v(\cdot - x_0)\|_{L^\infty(B_r(x_0))} \le \varepsilon r \quad \text{for some} \quad v \in \partial B_1,$$

   then $\partial\Omega_{u_n} = Reg\,(\partial\Omega_{u_n})$ in $B_{r/2}(x_0)$.

(b) **Uniform non-degeneracy.** *There are constants $\kappa > 0$ and $r_0 > 0$ such that the following holds: if $n \in \mathbb{N}$, $x_0 \in \partial\Omega_{u_n} \cap D$ and $r \in (0, r_0)$ are such that $B_r(x_0) \subset D$, then*

$$\|u_n\|_{L^\infty(B_r(x_0))} \ge \kappa r .$$

(c) **Uniform convergence.** *The sequence u_n converges locally uniformly in D to a (continuous and non-negative) function $u_0 : D \to \mathbb{R}$.*

Then, for every compact set $K \subset D$, the following claim does hold:

> *For every open set $U \subset D$ containing $\operatorname{Sing}(\partial\Omega_{u_0}) \cap K$,*
> *there exists $n_0 \in \mathbb{N}$ such that:*   (10.5)
> *$\operatorname{Sing}(\partial\Omega_{u_n}) \cap K \subset U$ for every $n \ge n_0$.*

In particular, for every $s > 0$,

$$\mathcal{H}^s_\infty\big(Sing\,(\partial\Omega_{u_0}) \cap K\big) \ge \limsup_{n\to\infty} \mathcal{H}^s_\infty\big(\operatorname{Sing}(\partial\Omega_{u_n}) \cap K\big). \tag{10.6}$$

Proof The semicontinuity of the Hausdorff measure (10.6) follows by (10.5) and the definition of $\mathcal{H}^s_\infty$. Thus, it is sufficient to prove (10.5). Arguing by contradiction, we suppose that there are a compact set $K \subset D$ and an open set $U \subset D$ such that

$$\operatorname{Sing}(\partial\Omega_{u_0}) \cap K \subset U,$$

but (up to extracting a subsequence of u_n) there is a sequence

$$x_n \in \operatorname{Sing}(\partial\Omega_{u_n}) \cap K \cap (\mathbb{R}^d \setminus U).$$

Up to extracting a further sequence we may assume that there is a point x_0 such that

$$x_0 \in K \cap (\mathbb{R}^d \setminus U) \qquad \text{and} \qquad x_0 = \lim_{n\to\infty} x_n.$$

We claim that $x_0 \in \partial\Omega_{u_0}$. Indeed, the uniform convergence of u_n implies that $u_0(x_0) = 0$. On the other hand, the non-degeneracy hypothesis (b) implies that, for every $r > 0$ small enough,

$$\|u_0\|_{L^\infty(B_r(x_0))} \ge \liminf_{n\to\infty} \Big(\|u_n\|_{L^\infty(B_r(x_0))} - \|u_n - u_0\|_{L^\infty(B_r(x_0))} \Big)$$

$$\ge \liminf_{n\to\infty} \|u_n\|_{L^\infty(B_{r/2}(x_n))} \ge \kappa \frac{r}{2},$$

which gives that $x_0 \in \partial\Omega_{u_0}$.

Now, we notice that, since U contains $Sing(\partial\Omega_{u_0}) \cap K$ and $x_0 \notin U$, we have that

$$x_0 \in Reg(\partial\Omega_{u_0}).$$

By definition of $Reg(\partial\Omega_{u_0})$, there is a sequence $r_n \to 0$ and a unit vector $\nu \in \partial B_1$ such that

$$\lim_{n\to\infty} \frac{1}{r_n} \|u_0 - h_\nu(\cdot - x_0)\|_{L^\infty(B_{r_n}(x_0))} = 0.$$

In particular, there exists $r \in (0, R)$ such that $B_r(x_0) \subset D$ and

$$\|u_0(x) - h_\nu(\cdot - x_0)\|_{L^\infty(B_r(x_0))} \le \frac{\varepsilon}{3} r.$$

By the continuity of u_0 and h_ν, we get that, for n large enough,

$$\|u_0 - h_\nu(\cdot - x_n)\|_{L^\infty(B_r(x_n))} \le \frac{2\varepsilon}{3} r.$$

Since, u_n converges to u_0 locally uniformly in D, we get that, for n large enough,

$$\|u_n - h_\nu(\cdot - x_n)\|_{L^\infty(B_r(x_n))} \le \varepsilon r.$$

Thus, (a) implies that $x_n \in \text{Reg}(\partial\Omega_{u_n})$, in contradiction with the initial assumption.

$\square$

10.3 Dimension Reduction

In this section, we study the singularities of the global one-homogeneous minimizers of $\mathcal{F}_\Lambda$. In particular, we prove Theorem 1.4 in the case when u is one-homogeneous. This (significant) simplification is essential for the proof of Theorem 1.4 since we already know that the blow-up limits of a local minimizer are global one-homogeneous minimizers and we will prove (see Lemma 10.7) that the dimension of the singular set of a blow-up limit does not decrease if we choose the free boundary point to have non-zero Hausdorff density (see Lemma 10.5).

Remark 10.8 (The Singular Set of a One-Homogeneous Function Is a Cone) Suppose that $z : \mathbb{R}^d \to \mathbb{R}$ is a non-negative one-homogeneous local minimizer of $\mathcal{F}_\Lambda$ in $\mathbb{R}^d$. Then, for any singular free boundary point $x_0 \in \text{Sing}(\partial\Omega_z) \setminus \{0\}$, we have that $\{tx_0 : t \in \mathbb{R}\} \subset \text{Sing}(\partial\Omega_z)$. This claim follows by the fact that $\text{Reg}(\partial\Omega_u)$ is a cone. and that

$$\text{Sing}(\partial\Omega_z) = \partial\Omega_z \setminus \text{Reg}(\partial\Omega_z).$$

Lemma 10.9 (Blow-Up Limits of One-Homogeneous Functions) *Let $z : \mathbb{R}^d \to \mathbb{R}$ be a one-homogeneous locally Lipschitz continuous function. Let $0 \ne x_0 \in \partial\Omega_z$.*

Let $r_n \to 0$ and z_{r_n,x_0} be a a blow-up sequence converging locally uniformly to a function $z_0 : \mathbb{R}^d \to \mathbb{R}$. Then z_0 is invariant in the direction x_0, that is,

$$z_0(x + tx_0) = z_0(x) \quad \text{for every} \quad x \in \mathbb{R} \quad \text{and every} \quad t \in \mathbb{R}.$$

Proof Let $t \in \mathbb{R}$ be fixed. Then, we have

$$z_0(x + tx_0) = \lim_{n \to \infty} z_{r_n,x_0}(x + tx_0) = \lim_{n \to \infty} \frac{1}{r_n} z\big(x_0 + r_n(x + tx_0)\big)$$

$$= \lim_{n \to \infty} \frac{1 + tr_n}{r_n} z\Big(x_0 + \frac{r_n}{1 + tr_n}x\Big) = \lim_{n \to \infty} \frac{1}{r_n} z\big(x_0 + r_n x\big) = z_0(x),$$

where the third inequality follows by the homogeneity of z and the fourth inequality follows by the Lipschitz continuity of z. Precisely, setting $L = \|\nabla z\|_{L^\infty(B_1(x_0))}$, we have

$$\Big| \frac{1 + tr_n}{r_n} z\Big(x_0 + \frac{r_n}{1 + tr_n}x\Big) - \frac{1}{r_n} z\big(x_0 + r_n x\big) \Big|$$

$$\leq t\,|z|\,\Big(x_0 + \frac{r_n}{1 + tr_n}x\Big) + \frac{1}{r_n} \Big| z\Big(x_0 + \frac{r_n}{1 + tr_n}x\Big) - z\big(x_0 + r_n x\big) \Big|$$

$$\leq t \frac{r_n L |x|}{1 + tr_n} + \frac{1}{r_n} \frac{tr_n^2 L |x|}{1 + tr_n},$$

which converges to zero as $n \to \infty$. $\square$

Lemma 10.10 (Translation Invariant Global Minimizers) *Let $u : \mathbb{R}^{d-1} \to \mathbb{R}$ be a non-negative function, $u \in H^1_{loc}(\mathbb{R}^{d-1})$ and let $\tilde{u} : \mathbb{R}^d \to \mathbb{R}$ be the function defined by*

$$\tilde{u}(x) = u(x') \quad \text{for every} \quad x = (x', x_d) \in \mathbb{R}^d.$$

Then, u a local minimizer of $\mathcal{F}_\Lambda$ in $\mathbb{R}^{d-1}$ if and only if $\tilde{u}$ a local minimizer of $\mathcal{F}_\Lambda$ in $\mathbb{R}^d$.

Proof Suppose first that $\tilde{u}$ is not a local minimizer of $\mathcal{F}_\Lambda$. Then, there is a function $\tilde{v} : \mathbb{R}^d \to \mathbb{R}$ such that $\tilde{u} = \tilde{v}$ outside the cylinder $\mathcal{C}_R := B'_R \times (-R, R) \subset \mathbb{R}^{d-1} \times \mathbb{R}$ and such that $\mathcal{F}_\Lambda(\tilde{u}, \mathcal{C}_R) > \mathcal{F}_\Lambda(\tilde{v}, \mathcal{C}_R)$.

$$\mathcal{F}_\Lambda(u, B_{R'}) = \int_{B'_R} |\nabla_{x'} u|^2 \, dx' + \Lambda \big| B'_R \cap \{u > 0\} \big|$$

$$= \frac{1}{2R} \left(\int_{\mathcal{C}_R} |\nabla \tilde{u}|^2 \, dx + \Lambda \big| \mathcal{C}_R \cap \{\tilde{u} > 0\} \big| \right) = \frac{1}{2R} \mathcal{F}_\Lambda(\tilde{u}, \mathcal{C}_R)$$

$$> \frac{1}{2R} \mathcal{F}_\Lambda(\tilde{v}, \mathcal{C}_R) = \frac{1}{2R} \left(\int_{\mathcal{C}_R} |\nabla \tilde{v}|^2 \, dx + \Lambda \left| \mathcal{C}_R \cap \{\tilde{v} > 0\} \right| \right)$$

$$\geq \frac{1}{2R} \int_{-R}^R \left(\int_{B'_R} |\nabla_{x'} \tilde{v}(x', x_d)|^2 \, dx' + \Lambda \left| B'_R \cap \{\tilde{v}(\cdot, x_d) > 0\} \right| \right) dx_d$$

$$\geq \int_{B'_R} |\nabla_{x'} \tilde{v}(x', t)|^2 \, dx' + \Lambda \left| B'_R \cap \{\tilde{v}(\cdot, t) > 0\} \right|,$$

for some $t \in (-R, R)$, which exists due to the mean-value theorem. Thus, also u is not a local minimizer of $\mathcal{F}_\Lambda$.

Conversely, suppose that u is not a local minimizer of $\mathcal{F}_\Lambda$. Then, there is a function $v : \mathbb{R}^{d-1} \to \mathbb{R}$ such that $u = v$ outside a ball $B'_R \subset \mathbb{R}^{d-1}$ and $\mathcal{F}_\Lambda(u, B'_R) > \mathcal{F}_\Lambda(v, B'_R)$. We now define the function

$$\tilde{v}(x', x_d) = v(x') \phi_t(x_d),$$

where for any $t > 0$, we define the function $\phi_t : \mathbb{R} \to [0, 1]$ as

$$\phi_t(x_d) := \begin{cases} 1 & \text{if } |x_d| \leq t, \\ 0 & \text{if } |x_d| \geq t + 1, \\ x_d + t + 1 & \text{if } -t - 1 \leq x_d \leq -t, \\ x_d - t & \text{if } t \leq x_d \leq t + 1. \end{cases}$$

Then,

$$|\nabla_x \tilde{v}|^2 \leq |\nabla_{x'} v|^2 + v^2 \mathbb{1}_{\mathcal{C}_{R,t+1} \setminus \mathcal{C}_{R,t}},$$

$$\left| \mathcal{C}_{R,t+1} \cap \{\tilde{v} > 0\} \right| = 2(t + 1) \left| B'_R \cap \{v > 0\} \right|,$$

where $\mathcal{C}_{R,t} := B'_R \times (-t, t)$. Thus, we have

$$\mathcal{F}_\Lambda(\tilde{v}, \mathcal{C}_{R,t+1}) = \int_{\mathcal{C}_{R,t+1}} |\nabla \tilde{v}|^2 \, dx + \Lambda \left| \mathcal{C}_{R,t+1} \cap \{\tilde{v} > 0\} \right|$$

$$\leq 2t \mathcal{F}_\Lambda(v, B'_R) + 2 \int_{B'_R} v^2 \, dx' + 2 \left| B'_R \cap \{v > 0\} \right|.$$

Choosing t large enough, we have that

$$2t \mathcal{F}_\Lambda(v, B'_R) + 2 \int_{B'_R} v^2 \, dx' + 2 \left| B'_R \cap \{v > 0\} \right| \leq 2t \mathcal{F}_\Lambda(u, B'_R).$$

Since,

$$\mathcal{F}_\Lambda(\tilde{u}, C_{R,t+1}) = 2(t+1)\mathcal{F}_\Lambda(u, B'_R),$$

we get that

$$\mathcal{F}_\Lambda(\tilde{v}, C_{R,t+1}) < \mathcal{F}_\Lambda(\tilde{u}, C_{R,t+1}),$$

which concludes the proof.   □

Lemma 10.11 (Singular One-Homogeneous Global Minimizers in $\mathbb{R}^{d^*}$) *Let $z :$ $\mathbb{R}^{d^*} \to \mathbb{R}$ be a non-negative one-homogeneous local minimizer of $\mathcal{F}_\Lambda$ in $\mathbb{R}^{d^*}$. Then, one of the following does hold:*

(1) $z(x) = \sqrt{\Lambda}\,(x \cdot v)$ for some $v \in \mathbb{R}^{d^*}$ (in this case $\mathrm{Sing}\,(\partial\Omega_z) = \emptyset$);
(2) $\mathrm{Sing}\,(\partial\Omega_z) = \{0\}$.

In other words,

$$\mathrm{Sing}\,(\partial\Omega_z) \setminus \{0\} = \emptyset.$$

In particular, this means that $\dim_{\mathcal{H}} \mathrm{Sing}\,(\partial\Omega_z) = 0$.

Proof Suppose that there is a point $x_0 \in \mathbb{R}^d \setminus \{0\}$ such that $x_0 \in \mathrm{Sing}\,(\partial\Omega_z)$. Then, by Remark 10.8 we have that $tx_0 \in \mathrm{Sing}\,(\partial\Omega_z)$ for every $t \in \mathbb{R}$. In particular, we can suppose that $|x_0| = 1$ and, without loss of generality, we set $x_0 = e_d$. Let now z_0 be a blow-up limit of z at x_0. Then, z_0 is a one-homogeneous local minimizer of $\mathcal{F}_\Lambda$. Moreover, by Lemma 10.9 we have that $z_0(x', t) = z_0(x', 0)$ for every $x' \in \mathbb{R}^{d-1}$. Now, Lemma 10.10 implies that the function $z'_0 := z_0(\cdot, 0) : \mathbb{R}^{d-1} \to \mathbb{R}$ is still a local minimizer of $\mathcal{F}_\Lambda$. Moreover, the origin $0' \in \mathbb{R}^{d-1}$ is a singular point for $\partial\Omega_{z'_0}$ in contradiction with the definition of d^*.   □

Lemma 10.12 (Dimension Reduction: Lemma II) *Suppose that $d \geq d^*$ and that $z : \mathbb{R}^d \to \mathbb{R}$ is a non-negative one-homogeneous local minimizer of $\mathcal{F}_\Lambda$ in $\mathbb{R}^d$. Then,*

$$\mathcal{H}^{d-d^*+s}\big(\mathrm{Sing}\,(\partial\Omega_z)\big) = 0 \qquad \text{for every} \qquad s > 0.$$

Proof Let $s > 0$ be fixed. The claim in the case $d = d^*$ follows by Lemma 10.11. We will prove the claim by induction. Indeed, suppose that the claim holds in dimension $d - 1$, with $d - 1 \geq d^*$, and let $z : \mathbb{R}^d \to \mathbb{R}$ be a non-negative one-homogeneous local minimizer. If such that $\mathcal{H}^{d-d^*+s}\big(\mathrm{Sing}\,(\partial\Omega_z)\big) > 0$, then, by Lemma 10.5, there is a point $x_0 \in \mathrm{Sing}\,(\partial\Omega_z)$, a constant $\varepsilon > 0$ and a sequence $r_n \to 0$ such that

$$\mathcal{H}^{d-d^*+s}\big(\mathrm{Sing}\,(\partial\Omega_z) \cap B_{r_n}(x_0)\big) \geq \varepsilon r_n^{d-d^*+s} \qquad \text{for every} \qquad n \in \mathbb{N},$$

which can be re-written as

$$\mathcal{H}^{d-d^*+s}\big(\mathrm{Sing}\,(\partial\Omega_{z_n})\cap B_1\big) \geq \varepsilon \qquad \text{for every} \qquad n \in \mathbb{N}, \qquad (10.7)$$

where we have set $z_n(x) := \frac{1}{r_n}z(x_0 + r_n x)$.

Without loss of generality, we can assume that $x_0 = e_d$. Now, up to a subsequence, z_n converges to a blow-up limit z_0 of z. By Lemma 10.9 and Lemma 10.10, we have that:

(1) $z_0(x', x_d) = z_0(x', 0)$ for every $x' \in \mathbb{R}^{d-1}$ and every $x_d \in \mathbb{R}$;
(2) $z_0' := z_0(\cdot, 0) : \mathbb{R}^{d-1} \to \mathbb{R}$ is one-homogeneous local minimizer of $\mathcal{F}_\Lambda$ in $\mathbb{R}^{d-1}$.

By hypothesis, we have that

$$\mathcal{H}^{d-1-d^*+s}\big(\mathrm{Sing}\,(\partial\Omega_{z_0'})\big) = 0.$$

The translation invariance of z_0 now implies that

$$\mathrm{Sing}\,(\partial\Omega_{z_0}) = \mathrm{Sing}\,(\partial\Omega_{z_0'}) \times \mathbb{R},$$

so, Lemma 10.6 gives

$$\mathcal{H}^{d-d^*+s}\big(\mathrm{Sing}\,(\partial\Omega_{z_0})\big) = 0,$$

which is a contradiction with (10.6) of Lemma 10.7 and (10.7).   □

10.4   Proof of Theorem 1.4

In this section, we will give an estimate on the dimension of the singular set. The result is more general and applies to different situations, for instance to almost-minimizers and measure-constrained minimizers.

Proposition 10.13 (Dimension of the Singular Set) *Let $D \subset \mathbb{R}^d$ be a bounded open set and $u : D \to \mathbb{R}$ a continuous non-negative function. Let the regular and singular sets $\mathrm{Reg}(\partial\Omega_u)$ and $\mathrm{Sing}(\partial\Omega_u)$ of the free boundary $\partial\Omega_u \cap D$ be defined as in the beginning of Sect. 10.2. Suppose that u satisfies the following hypotheses:*

(a) ε-regularity. There are constants $\varepsilon > 0$ and $R > 0$ such that the following holds:

If $x_0 \in \partial\Omega_u \cap D$ and $r \in (0, R)$ are such that $B_r(x_0) \subset D$ and

$$\|u(x) - \sqrt{\Lambda}\,((x-x_0)\cdot v)_+\|_{L_x^\infty(B_r(x_0))} \leq \varepsilon r \quad \text{for some} \quad v \in \partial B_1, \quad (10.8)$$

then $\partial\Omega_u = \mathrm{Reg}\,(\partial\Omega_u)$ in $B_{r/2}(x_0)$.

(b) **Non-degeneracy.** *There are constants* $\kappa > 0$ *and* $r_0 > 0$ *such that the following holds: if* $n \in \mathbb{N}$, $x_0 \in \partial\Omega_u \cap D$ *and* $r \in (0, r_0)$ *are such that* $B_r(x_0) \subset D$, *then*

$$\|u\|_{L^\infty(B_r(x_0))} \geq \kappa r .$$

(c) **Convergence of the blow-up sequences.** *Every blow-up sequence*

$$u_{r_n, x_0}(x) = \frac{1}{r_n} u(x_0 + r_n x),$$

with $x_0 \in \partial\Omega_u \cap D$ *and* $r_n \to 0$, *admits a subsequence that converges locally uniformly to a blow-up limit* $u_0 : \mathbb{R}^d \to \mathbb{R}$.

(d) **Homogeneity and minimality of the blow-up limits.** *Every blow-up limit of* u *is a one-homogeneous global minimizer of* $\mathcal{F}_\Lambda$ *in* $\mathbb{R}^d$.

Then,

  (i) *if* $d < d^*$, *then* $\mathrm{Sing}\,(\partial\Omega_u)$ *is empty;*
  (ii) *if* $d = d^*$, *then* $\mathrm{Sing}\,(\partial\Omega_u)$ *is locally finite;*
  (iii) *if* $d > d^*$, *then* $\dim_{\mathcal{H}} \mathrm{Sing}\,(\partial\Omega_u) \leq d - d^*$.

Proof Suppose first that $d < d^*$. Let $x_0 \in \partial\Omega_u \cap D$ and let $r_n \to 0$ be a infinitesimal sequence such that u_{r_n, x_0} converges locally uniformly to a blow-up limit u_0 (such a sequence exists by the hypothesis (b)). By (c), u_0 is a one-homogeneous local minimizer of $\mathcal{F}_\Lambda$ in $\mathbb{R}^d$. By definition of d^*, we get that $\mathrm{Sing}\,(\partial\Omega_{u_0}) = \emptyset$. This means that every blow-up limit of u_0 is of the form $\sqrt{\Lambda}\,(x \cdot \nu)_+$, for some $\nu \in \partial B_1$. In particular, it holds for every blow-up limit in zero. Since u_0 is one-homogeneous, the blow-up of u_0 in zero is u_0 itself and so,

$$u_0(x) = \sqrt{\Lambda}\,(x \cdot \nu)_+ \qquad \text{for some} \qquad \nu \in \partial B_1.$$

Thus, for n large enough, we get that

$$\|u_{r_n, x_0}(x) - \sqrt{\Lambda}\,(x \cdot \nu)_+\|_{L^\infty_x(B_1)} \leq \varepsilon,$$

which, by the definition of u_{r_n, x_0} gives precisely (10.8). Thus, by (a), we get that x_0 is a regular point, $x_0 \in \mathrm{Reg}\,(\partial\Omega_u)$. Since x_0 is arbitrary, we conclude that $\mathrm{Sing}\,(\partial\Omega_u) = \emptyset$.

Let now $d = d^*$. Suppose by contradiction that there is a sequence of points $x_n \in \mathrm{Sing}\,(\partial\Omega_u)$ converging to a point $x_0 \in D \cap \mathrm{Sing}\,(\partial\Omega_u)$. Let $r_n := |x_n - x_0|$. Up to extracting a subsequence, we can assume that the blow-up sequence $u_n := u_{r_n, x_0}$ converges to a blow-up limit $u_0 : \mathbb{R}^d \to \mathbb{R}$. By (c), u_0 is a one-homogeneous local minimizer of $\mathcal{F}_\Lambda$ in $\mathbb{R}^d$. On the other hand, notice that for every $n > 0$ the point $\xi_n = \frac{x_n - x_0}{r_n} \in \partial B_1$ is a singular point for u_n. Up to extracting a subsequence, we may assume that ξ_n converges to a point $\xi_0 \in \partial B_1$. By Lemma 10.7, we get that $\xi_0 \in \mathrm{Sing}\,(\partial\Omega_{u_0})$, in contradiction with Lemma 10.11.

Finally, we consider the case $d > d^*$. Let $s > 0$ be fixed. We will prove that $\mathcal{H}^{d-d^*+s}\big(\mathrm{Sing}\,(\partial\Omega_u)\big) = 0$. Suppose that this is not the case and $\mathcal{H}^{d-d^*+s}\big(\mathrm{Sing}\,(\partial\Omega_u)\big) > 0$. By Lemma 10.5 we have that there is a point $x_0 \in \mathrm{Sing}\,(\partial\Omega_u)$ and a sequence $r_n \to 0$ such that

$$\mathcal{H}^{d-d^*+s}\big(\mathrm{Sing}\,(\partial\Omega_u) \cap B_{r_n}(x_0)\big) \geq \varepsilon r_n^{d-d^*+s}.$$

Taking, $u_n = u_{r_n,x_0}$, we get that

$$\mathcal{H}^{d-d^*+s}\big(\mathrm{Sing}\,(\partial\Omega_{u_n}) \cap B_1\big) \geq \varepsilon.$$

Using (b), we can suppose that, up to extracting a subsequence, u_n converges to a blow-up limit u_0. By (c), u_0 is a one-homogeneous minimizer of $\mathcal{F}_\Lambda$ in $\mathbb{R}^d$. Now, Lemma 10.7, we get that $\mathcal{H}^{d-d^*+s}\big(\mathrm{Sing}\,(\partial\Omega_{u_0}) \cap B_1\big) \geq \varepsilon$, which is in contradiction with Lemma 10.12.

$\square$

Chapter 11
Regularity of the Free Boundary for Measure Constrained Minimizers

Let D be a connected bounded open set in $\mathbb{R}^d$ and let $v \in H^1(D)$ be a given non-negative function. This chapter is dedicated to the problem

$$\min \left\{ \mathcal{F}_0(u, D) \ : \ u \in H^1(D), \ u - v \in H^1_0(D), \ |\Omega_u \cap D| = m \right\}, \qquad (11.1)$$

where $m \in (0, |D|)$ is a fixed constant and we recall that

$$\mathcal{F}_0(u, D) = \int_D |\nabla u|^2 \, dx.$$

In this chapter, we give the main steps of the proof of Theorem 1.9.

- Section 11.1. *Existence of minimizers.*
  In this section, we prove that (11.1) admits a solution and that every solution is a non-negative subharmonic function (see Proposition 11.1).
- Section 11.2. *Euler-Lagrange equations.*
  In this section, we prove that if u is a solution to (11.1), then there exists a Lagrange multiplier $\Lambda \geq 0$ such that the first variation of $\mathcal{F}_\Lambda$ vanishes, that is,

$$\delta \mathcal{F}_\Lambda(u, D)[\xi] = 0 \quad \text{for every} \quad \xi \in C^\infty_c(D; \mathbb{R}^d).$$

- Section 11.3. *Strict positivity of the Lagrange multiplier.*
  In this section we prove that $\Lambda > 0$.
- Section 11.4. *Convergence of the Lagrange multipliers.*
  In this section, we prove a technical lemma, that we will use several times in the next section. Roughly speaing, we show that if u_n is a sequence of solutions converging to a solution u, then also the sequence of Lagrange multipliers converge to the Lagrange multipliers of u.
- Section 11.5. *Almost optimality of u at small scales.*

© The Author(s) 2023
B. Velichkov, *Regularity of the One-phase Free Boundaries*,
Lecture Notes of the Unione Matematica Italiana 28,
https://doi.org/10.1007/978-3-031-13238-4_11

In this section, we show that if u is a solution to (11.1), then it minimizes $\mathcal{F}_\Lambda$ in every ball B_r, up to an error that depends on the radius r and vanishes as $r \to 0$. At this point, the regularity of the free boundary (Theorem 1.9) follows by the same arguments that we used for Theorem 1.2 and Theorem 1.4, the necessary modifications being pointed out in the sketch of the proof given in the introduction.

11.1   Existence of Minimizers

In this section we prove that there is a solution to the problem (11.1). This follows by a standard argument which can be divided in two steps. We will first show that there is a solution u to the auxiliary problem

$$\min \left\{ \mathcal{F}_0(u, D) \; : \; u \in H^1(D), \; u - v \in H^1_0(D), \; |\Omega_u^+ \cap D| \leq m \right\}, \tag{11.2}$$

where for simplicity we set

$$\Omega_u^+ = \Omega_{u_+} = \{u > 0\}.$$

Then we will prove that the constraint is saturated, that is, every solution u of (11.2) is such that $|\Omega_u| = |\Omega_u^+| = m$. We give the details in the following proposition.

Proposition 11.1 (Existence of Minimizers) *Let D be a connected bounded open set in $\mathbb{R}^d$, $v \in H^1(D)$ be a non-negative function and $0 < m < |D|$. Then,*

(i) *there is a solution to the variational problem (11.1);*
(ii) *the function $u \in H^1(D)$ is a solution to (11.1) if and only if it is a solution to (11.2);*
(iii) *every solution (to (11.1) and (11.2)) is a non-negative subharmonic function in D and, in particular, is defined at every point of D.*

Proof We will proceed in several steps.
Step 1. There is a solution to the auxiliary problem (11.2). This follows by a direct argument. Indeed, let u_n be a minimizing sequence for (11.2), that is, $u_n - v \in H^1_0(D)$, $|\Omega_u^+ \cap D| \leq m$ and

$$\lim_{n \to \infty} \mathcal{F}_0(u_n, D) = \inf \left\{ \mathcal{F}_0(u, D) \; : \; u \in H^1(D), \; u - v \in H^1_0(D), \; |\Omega_u^+ \cap D| \leq m \right\}.$$

Then, up to a subsequence, u_n converges weakly in $H^1(D)$, strongly in $L^2(D)$ and pointwise a.e. in D to a function $u_\infty \in H^1(D)$ such that $u_\infty - v \in H^1_0(D)$. Then, we have

$$\mathcal{F}_0(u_\infty, D) \leq \liminf_{n \to \infty} \mathcal{F}_0(u_n, D),$$

and, by the pointwise convergence of u_n to u_∞,

$$\mathbb{1}_{\{u_\infty>0\}} \le \liminf_{n\to\infty} \mathbb{1}_{\{u_n>0\}} \qquad \text{and} \qquad |\{u_\infty > 0\}| \le \liminf_{n\to\infty} |\{u_n > 0\}| \le m,$$

which means that u_∞ is a solution to (11.2).

Step 2. Every solution u to (11.2) *is non-negative.* Indeed, this follows simply by the fact that if $u = u_+ - u_-$ is a solution to (11.2), then the function u_+ still satisfies the constraints $u_+ - v \in H_0^1(\Omega)$ and $|\Omega_{u_+}| \le m$, and we have

$$\mathcal{F}_0(u, D) = \mathcal{F}_0(u_+, D) + \mathcal{F}_0(u_-, D) \le \mathcal{F}_0(u_+, D),$$

with an equality if and only if $u_- \equiv 0$.

Step 3. Every solution u to (11.2) *is subharmonic.* This follows by the fact that

$$\mathcal{F}_0(u, D) \le \mathcal{F}_0(\varphi, D),$$

for every $\varphi \le u$ with the same boundary values as u. In particular, this means that u is defined pointwise. In fact, we simply consider the representative of u defined as

$$u(x_0) := \lim_{r\to 0} \fint_{B_r(x_0)} u(x)\, dx = \lim_{r\to 0} \fint_{\partial B_r(x_0)} u\, d\mathcal{H}^{d-1}.$$

Step 4. Every solution u to (11.2) *satisfies the condition* $|\Omega_u| = m$. Indeed, suppose that this is not the case. Let $r_0 > 0$ be such that $|B_{r_0}| \le m - |\Omega_u|$. Take $x_0 \in D$ and $r < \min\{r_0, \text{dist}(x_0, \partial D)\}$. Let h be the harmonic extension of u in $B_r(x_0)$, that is, h is a solution of the PDE

$$\Delta h = 0 \quad \text{in} \quad B_r(x_0), \qquad h = u \quad \text{on} \quad \partial B_r(x_0).$$

Then, let $\widetilde{u}$ be the competitor defined as

$$\widetilde{u} = h \quad \text{in} \quad B_r(x_0), \qquad \widetilde{u} = u \quad \text{in} \quad D \setminus B_r(x_0).$$

Then, $|\Omega_{\widetilde{u}}| \le m$ and so, the optimality of u gives

$$0 \ge \mathcal{F}_0(u, D) - \mathcal{F}_0(\widetilde{u}, D) = \int_{B_r(x_0)} |\nabla h|^2\, dx - \int_{B_r(x_0)} |\nabla u|^2\, dx = \int_{B_r(x_0)} |\nabla(u-h)|^2\, dx,$$

which means that $h = u$ in $B_r(x_0)$. In particular, we get that the set $\{u > 0\}$ is open: if $u(x_0) > 0$, then $\fint_{B_r(x_0)} u(x)\, dx > 0$ for some r small enough, but then $u > 0$ in $B_r(x_0)$ because it coincides with its (non-zero) harmonic extension. On the other hand $\{u > 0\}$ is also closed. Indeed, if there is a sequence of points x_n converging to x_0 such that $u(x_n) > 0$, then the harmonic extension of u in $B_r(x_0)$ is non-zero, so it is strictly positive, and so, $u(x_0) > 0$. Since D is connected, this means that $\{u > 0\} = D$, which is a contradiction with the fact that $|\Omega_u| \le m < |D|$.

Step 5. $u \in H^1(D)$ is a solution to (11.2) if and only if it is a solution to (11.1). This is a trivial consequence of Step 4.   □

In the rest of this section, without loss of generality, we will only consider functions $u \in H^1(D)$, which are non-negative and satisfy the following optimality condition:

$$\mathcal{F}_0(u, D) \leq \mathcal{F}_0(v, D) \quad \text{for every} \quad v \in H^1(D) \quad \text{such that} \quad \begin{cases} v - u \in H_0^1(D), \\ |\Omega_u| = |\Omega_v|. \end{cases}$$

(11.3)

11.2 Euler-Lagrange Equation

In this section, we will prove the existence of a Lagrange multiplier for functions satisfying (11.3). We will follow step-by-step the proof from [46]. Our main result is the following.

Proposition 11.2 (Euler-Lagrange Equation) *Let $D \subset \mathbb{R}^d$ be a connected bounded open set and let the non-negative function $u : D \to \mathbb{R}$ be a solution of (11.3). Then, there is a constant $\Lambda_u > 0$ such that*

$$\delta\mathcal{F}_0(u, D)[\xi] + \Lambda_u \int_{\Omega_u} div\, \xi \, dx = 0 \quad \text{for every} \quad \xi \in C_c^\infty(D; \mathbb{R}^d).$$

(11.4)

We start with the following lemma.

Lemma 11.3 (Variation of the Measure) *Let D be a connected open set in $\mathbb{R}^d$ and let $\Omega \subset D$ be a Lebesgue measurable set such that $0 < |\Omega| < |D|$. Then, there is a smooth vector field $\xi \in C_c^\infty(D; \mathbb{R}^d)$ such that*

$$\int_\Omega div\, \xi \, dx = 1.$$

Proof Assume, by contradiction, that we have

$$\int_\Omega div\, \xi \, dx = 0 \quad \text{for every} \quad \xi \in C_c^\infty(D; \mathbb{R}^d).$$

(11.5)

In particular, for every ball $B_\rho(x_0) \subset D$, we may choose ξ to be the vector field

$$\xi(x) = (x - x_0)\phi_\varepsilon(x),$$

where

$$0 \leq \phi_\varepsilon \leq 1 \quad \text{and} \quad |\nabla \phi_\varepsilon| \leq \frac{1 + \varepsilon\rho}{\varepsilon\rho} \quad \text{in} \quad B_\rho(x_0),$$

$$\phi_\varepsilon = 1 \quad \text{in} \quad B_{\rho(1-\varepsilon)}(x_0) \qquad \text{and} \qquad \phi_\varepsilon = 0 \quad \text{on} \quad \partial B_\rho(x_0).$$

By (11.5), we have

$$0 = \int_\Omega \operatorname{div} \xi \, dx = \int_\Omega \left(d\phi_\varepsilon(x) + (x - x_0) \cdot \nabla \phi_\varepsilon(x) \right) dx.$$

Passing to the limit as $\varepsilon \to 0$, we obtain

$$d|\Omega \cap B_\rho(x_0)| - \rho \, \mathcal{H}^{d-1}\left(\Omega \cap \partial B_\rho(x_0) \right) = 0.$$

In particular, we get that

$$\frac{\partial}{\partial \rho} \left(\frac{|\Omega \cap B_\rho(x_0)|}{\rho^d} \right) = 0,$$

which means that the function $\rho \mapsto \rho^{-d} |\Omega \cap B_\rho(x_0)|$ is constant. In particular, if $x_0 \in B_r$ is a point of zero Lebesgue density for Ω, then Ω has zero Lebesgue measure in a neighborhood of x_0. Precisely, setting $r(x) := \operatorname{dist}(x, \mathbb{R}^d \setminus D)$ we have that

$$\text{If} \quad x_0 \in \Omega^{(0)}, \quad \text{then} \quad |\Omega \cap B_{r(x_0)}(x_0)| = 0. \tag{11.6}$$

Now, notice that (11.6) is both an open and a closed subset of D. Since, by hypothesis, D is connected, we have that $\Omega^{(0)} = D$ or $\Omega^{(0)} = \emptyset$, which concludes the proof. $\qquad \square$

Proof of Proposition 11.2 Let $\xi \in C_c^\infty(D; \mathbb{R}^d)$. Using the notations from Lemma 9.5, for any (small enough) $t \in \mathbb{R}$, we set

$$\Psi_t = Id + t\xi, \qquad \Phi_t = \Psi_t^{-1} \quad \text{and} \quad u_t := u \circ \Psi_t.$$

By Lemma 9.5, we have that

$$|\Omega_{u_t}| = |\Omega_u| + t \int_{\Omega_u} \operatorname{div} \xi \, dx + o(t) \qquad \text{and}$$

$$\mathcal{F}_0(u_t, D) = \mathcal{F}_0(u, D) + t \, \delta \mathcal{F}_0(u, D)[\xi] + o(t).$$

Now, let the vector field $\xi_0 \in C_c^\infty(D; \mathbb{R}^d)$ be such that

$$\int_{\Omega_u} \operatorname{div} \xi_0 \, dx = 1 .$$

We are now going to prove that (11.4) holds with

$$\Lambda_u := -\delta \mathcal{F}_0(u, D)[\xi_0].$$

We fix $\xi \in C_c^\infty(D; \mathbb{R}^d)$ and we consider two cases.

Case 1. $\displaystyle\int_{\Omega_u} \operatorname{div} \xi \, dx = 0.$

Let $\xi_1 := \xi - \eta \xi_0$, where $\eta > 0$ is a real constant. Then, we have

$$\int_{\Omega_u} \operatorname{div} \xi_1 \, dx = -\eta.$$

Setting $u_t = u \circ \Phi_t$, where $\Phi_t := (Id + t\xi_1)^{-1}$, we have that, for $t > 0$ small enough,

$$u_t \in H_0^1(D) \qquad \text{and} \qquad |\Omega_{u_t}| \le |\Omega_u|.$$

By Proposition 11.1 (see Step 5 of the proof), we have that

$$\mathcal{F}_0(u, D) \le \mathcal{F}_0(u_t, D).$$

Taking the derivative at $t = 0$, we obtain

$$\delta \mathcal{F}_0(u, D)[\xi_1] \ge 0,$$

which can be re-written as

$$\delta \mathcal{F}_0(u, D)[\xi] \ge \eta \, \delta \mathcal{F}_0(u, D)[\xi_0].$$

Since η is arbitrary, we can deduce that

$$\delta \mathcal{F}_0(u, D)[\xi] \ge 0.$$

Finally, repeating the same argument for $-\xi$ instead of ξ, we obtain that

$$\delta \mathcal{F}_0(u, D)[\xi] = 0,$$

which concludes the proof of (11.4) in this case.

Case 2. $\displaystyle\int_{\Omega_u} \operatorname{div} \xi \, dx \neq 0.$

Let $\xi_2 := \xi - \xi_0 \displaystyle\int_{\Omega_u} \operatorname{div} \xi \, dx$. Then $\displaystyle\int_{\Omega_u} \operatorname{div} \xi_2 \, dx = 0$ and, by Case 1, we obtain

$$0 = \delta \mathcal{F}_0(u, D)[\xi_2] = \delta \mathcal{F}_0(u, D)[\xi] - \delta \mathcal{F}_0(u, D)[\xi_0] \int_{\Omega_u} \operatorname{div} \xi \, dx$$

$$= \delta \mathcal{F}_0(u, D)[\xi] + \Lambda_u \int_{\Omega_u} \operatorname{div} \xi \, dx,$$

which concludes the proof of (11.4).

It only remains to prove that $\Lambda_u \geq 0$. Indeed, let $u_t = u \circ \Phi_t$, where $\Phi_t = (Id - t\xi_0)^{-1}$. For $t > 0$ small enough, we have that $|\Omega_{u_t}| \leq |\Omega_u|$. We reason as in Case 1. By Proposition 11.1, we get that $\mathcal{F}_0(u, D) \leq \mathcal{F}_0(u_t, D)$. Then, taking the derivative at $t = 0$, we deduce

$$\Lambda_u := \delta \mathcal{F}_0(u, D)[-\xi_0] \geq 0.$$

The strict positivity of Λ_u is more involved and follows by Proposition 11.4, which we prove in the next subsection. □

11.3   Strict Positivity of the Lagrange Multiplier

In this section we prove that the Lagrange multiplier from Proposition 11.2 is strictly positive. Precisely, we will show that a function, which is critical for the functional $\mathcal{F}_0$ has a monotone Almgren frequency function $N(r)$. Now, the monotonicity of the frequency function implies that u cannot decay too fast around the free boundary points. On the other hand, if u is harmonic in Ω_u, then we can use a Caccioppoli inequality to show that if the Lebesgue density of Ω_u is too small, then the decay of u on the balls of radius r should be very fast. The combination of these two estimates implies that the Lebesgue density of Ω_u should be bounded from below at every point of D. In particular, there cannot be points of zero Lebesgue density for Ω_u in D. The results from this subsection come directly from [46], but this unique-continuation argument goes back to the work of Garofalo and Lin [34]. The main result of this subsection is the following.

Proposition 11.4 *Let D be a connected open set in $\mathbb{R}^d$. Suppose that $u \in H^1(D)$ is a non-identically-zero function such that*

(a) u is a solution of the equation

$$\Delta u = 0 \quad in \quad \Omega_u = \{u \neq 0\};$$

(b) u satisfies the extremality condition

$$\delta \mathcal{F}_0(u)[\xi] = 0 \quad \text{for every} \quad \xi \in C_c^\infty(D; \mathbb{R}^d),$$

where $\delta \mathcal{F}_0(u)[\xi]$ is the first variation of $\mathcal{F}_0$ in the direction ξ and is given by

$$\delta \mathcal{F}_0(u)[\xi] := \int_D \left[2\nabla u \cdot D\xi(\nabla u) - |\nabla u|^2 \text{div}\,\xi \right] dx. \tag{11.7}$$

Then, $|D \setminus \Omega_u| = 0$.

Remark 11.5 It is sufficient to prove Proposition 11.4 in the case $u \geq 0$. In fact, if $u : D \to \mathbb{R}$ satisfies the hypotheses (a) and (b) above, then the function $|u| : D \to \mathbb{R}$ satisfies the same hypotheses.

In the proof of Proposition 11.4 we will use the following Faber-Krahn-type inequality, which was first proved in [10] (we report here the original proof).

Lemma 11.6 (A Faber-Krahn Inequality, [10]) *There is a dimensional constant $C_d > 0$ such that for every ball $B_r \subset \mathbb{R}^d$ and every function $v \in H^1(B_r)$ satisfying*

$$\frac{|\Omega_v \cap B_r|}{|B_r|} \leq \frac{1}{2},$$

we have the inequality

$$\int_{B_r} v^2\, dx \leq C_d\, r^2 \left(\frac{|\Omega_v \cap B_r|}{|B_r|} \right)^{2/d} \int_{B_r} |\nabla v|^2\, dx, \tag{11.8}$$

where we recall that $\Omega_v = \{v \neq 0\}$.

Proof We first notice that:

- We can assume that v is non-negative in B_r. In fact, for every $v \in H^1(B_r)$, we have that $|v| \in H^1(B_r)$ and the following identities do hold:

$$\Omega_v = \Omega_{|v|}, \qquad v^2 = |v|^2 \qquad \text{and} \qquad |\nabla v|^2 = |\nabla |v||^2.$$

- We can assume that $r = 1$. Indeed, setting $v_r(x) = v(rx)$, we have that

$$|\Omega_v \cap B_r| = r^d |\Omega_{v_r} \cap B_1|, \qquad \int_{B_r} v^2\, dx = r^d \int_{B_1} v_r^2\, dx,$$

$$\int_{B_r} |\nabla v|^2\, dx = r^{d-2} \int_{B_1} |\nabla v_r|^2\, dx.$$

We now proceed with the proof of (11.8) in the case $r = 1$ and $v \geq 0$ on B_1.

Step 1. We claim that there is a dimensional constant $C_{iso} > 0$ such that

$$|\Omega|^{\frac{d-1}{d}} \le C_{iso}\, Per(\Omega; B_r) \quad \text{for every} \quad \Omega \subset B_r \quad \text{with} \quad |\Omega| \le \frac{1}{2}|B_r|, \quad (11.9)$$

where $Per(\Omega; B_r)$ is the relative perimeter in the sense of De Giorgi. The claim follows by a standard compactness argument.

Step 2. Let $n \in \mathbb{N}$ and let $D_n = \{x \cdot v_1 > 0\} \cap \{x \cdot v_2 > 0\}$ be the unbounded domain formed by the intersection of the two hyperplanes $\{x \cdot v_1 = 0\}$ and $\{x \cdot v_2 = 0\}$ forming (an interior) angle $2\pi/n$. We claim that, for every $\Omega \subset B_1$ satisfying $|\Omega| \le \frac{\omega_d}{2}$, there is a radius $\rho > 0$ such that

$$|B_\rho \cap D_n| = |\Omega| \quad \text{and} \quad Per(B_\rho \cap D_n; D_n) \le Per(\Omega; B_1). \quad (11.10)$$

Indeed, for every Ω, there is a unique $\rho > 0$ such that $|B_\rho \cap D_n| = |\Omega|$. We set $\Omega^* := B_\rho \cap D_n$. Then, we have

$$|\Omega^*|^{(d-1)/d} = n^{-(d-1)/d}|B_\rho|^{(d-1)/d} = \frac{n^{-(d-1)/d}}{d\omega_d^{1/d}}\, Per(B_\rho) = \frac{n^{1/d}}{d\omega_d^{1/d}}\, Per(\Omega^*; D_n).$$

Now, the isoperimetric inequality (11.9) implies

$$Per(\Omega; D) \ge C_{iso}^{-1}|\Omega|^{(d-1)/d} = C_{iso}^{-1}|\Omega^*|^{(d-1)/d} = C_{iso}^{-1}\frac{n^{1/d}}{d\omega_d^{1/d}}\, Per(\Omega^*; D_n).$$

Taking n large enough, such that $n^{1/d} \ge C_{iso}\, d\omega_d^{1/d}$, we get $P(\Omega; D) \ge Per(\Omega^*; D_n)$, which proves (11.10).

Step 3. For every non-negative function $v \in H^1(B_1)$ we define the symmetrized function $v_* \in H^1(D_n)$ obtained through the symmetrization of each level of v, that is,

$$\{v_* > t\} = \{v > t\}^* \quad \text{for every} \quad t \ge 0.$$

We claim that

$$\int_{D_n} v_*^2\, dx = \int_{B_1} v^2\, dx \quad \text{and} \quad \int_{D_n} |\nabla v_*|^2\, dx \le \int_{B_1} |\nabla v|^2\, dx. \quad (11.11)$$

The first part of (11.11) follows by the area formula

$$\int_{D_n} v_*^2\, dx = \int_0^\infty t|\{v_* > t\}|\, dt = \int_0^\infty t|\{v > t\}|\, dt = \int_{B_1} v^2\, dx,$$

while for the second part we will use the co-area formula. Indeed, setting

$$f(t) := |\{v > t\}| = |\{v^* > t\}|,$$

we have

$$
\begin{aligned}
\int_{B_1} |\nabla v|^2 \, dx &= \int_0^{+\infty} \left(\int_{\{v=t\}} |\nabla v| \, d\mathcal{H}^{d-1} \right) dt \\
&\geq \int_0^{+\infty} \left(\int_{\{v=t\}} |\nabla v|^{-1} \, d\mathcal{H}^{d-1} \right)^{-1} \left(\mathcal{H}^{d-1}(\{v = t\} \cap D) \right)^2 dt \\
&= \int_0^{+\infty} |f'(t)|^{-1} \left(\mathcal{H}^{d-1}(\{v = t\} \cap B_1) \right)^2 dt \\
&\geq \int_0^{+\infty} |f'(t)|^{-1} \left(\mathcal{H}^{d-1}(\{v_* = t\} \cap D_n) \right)^2 dt \\
&= \int_0^{+\infty} \left(\int_{\{v_*=t\}} |\nabla v_*|^{-1} \, d\mathcal{H}^{d-1} \right)^{-1} \left(\mathcal{H}^{d-1}(\{v_* = t\} \cap D_n) \right)^2 dt \\
&= \int_0^{+\infty} \left(\int_{\{v_*=t\}} |\nabla v_*| \, d\mathcal{H}^1 \right) dt = \int_{D_n} |\nabla v_*|^2 \, dx,
\end{aligned}
$$

where the first inequality follows by Cauchy-Schwartz and the second one is a consequence of (11.10).

Step 4. We claim that there is a constant $C_{d,n} > 0$, depending only on d and n, such that

$$\int_{D_n} v_*^2 \, dx \leq C_{d,n} \, |\{v_* > 0\}|^{2/d} \int_{D_n} |\nabla v_*|^2 \, dx. \tag{11.12}$$

Let $\tilde{v}_* : \mathbb{R}^d \to \mathbb{R}$ be the radially decreasing function defined by

$$\tilde{v}_*(x) = v_*(y),$$

where $y \in D_n$ is any point such that $|x| = |y|$. By the classical Faber-Krahn inequality in $\mathbb{R}^d$, there is a dimensional constant C_d such that

$$\int_{\mathbb{R}^d} \tilde{v}_*^2 \, dx \leq C_d \, |\{\tilde{v}_* > 0\}|^{2/d} \int_{\mathbb{R}^d} |\nabla \tilde{v}_*|^2 \, dx.$$

which gives that

$$\int_{D_n} v_*^2 \, dx \leq C_d \, n^{2/d} |\{v_* > 0\}|^{2/d} \int_{D_n} |\nabla v_*|^2 \, dx,$$

which is precisely (11.12). This, together with (11.11), concludes the proof.   □

In the next lemma, we prove that the Almgren frequency function is monotone. This is a classical result, which was first proved by Almgren [2].

Lemma 11.7 (Almgren Monotonicity Formula) *Let $u \in H^1(B_R)$. For $r \in (0, R]$, we define*

$$H(r) := \int_{\partial B_r} u^2 \, d\mathcal{H}^{d-1} \quad and \quad D(r) := \int_{B_r} |\nabla u|^2 \, dx,$$

and, if $H(r) \neq 0$, we define the Almgren frequency function

$$N(r) := \frac{r D(r)}{H(r)}.$$

Suppose that u is a solution of the equation

$$\Delta u = 0 \quad in \quad \Omega_u = \{u \neq 0\};$$

and satisfies the extremality condition

$$\delta \mathcal{F}_0(u)[\xi] = 0 \quad for \; every \quad \xi \in C_c^\infty(B_R; \mathbb{R}^d),$$

where $\delta \mathcal{F}_0(u)[\xi]$ is given by (11.7). If, moreover, $H > 0$ on the interval $(a, b) \subset (0, R)$, then the frequency function N is non-decreasing on (a, b).

Proof We first calculate the derivative of H

$$H'(r) = \frac{d-1}{r} H(r) + r^{d-1} \frac{\partial}{\partial r} \int_{\partial B_1} u^2(rx) \, d\mathcal{H}^{d-1}(x)$$

$$= \frac{d-1}{r} H(r) + 2 \int_{\partial B_r} u \frac{\partial u}{\partial n} d\mathcal{H}^{d-1} = \frac{d-1}{r} H(r) + 2 \int_{B_r} |\nabla u|^2 \, dx,$$

which we rewrite as

$$H'(r) = \frac{d-1}{r} H(r) + 2D(r). \tag{11.13}$$

Next, we notice that the extremality condition $\delta \mathcal{F}_0(u) = 0$ gives that the following equipartition of the energy does hold:

$$0 = -(d-2)\int_{B_r} |\nabla u|^2 \, dx + r \int_{\partial B_r} |\nabla u|^2 \, d\mathcal{H}^{d-1} - 2r \int_{\partial B_r} \left(\frac{\partial u}{\partial n}\right)^2 \, d\mathcal{H}^{d-1},$$

which can be rewritten as

$$-(d-2)D(r) + rD'(r) = 2r \int_{\partial B_r} \left(\frac{\partial u}{\partial n}\right)^2 \, d\mathcal{H}^{d-1}.$$

We now compute the derivative of N.

$$N'(r) = \frac{D(r)H(r) + rD'(r)H(r) - rD(r)H'(r)}{H^2(r)}$$

$$= \frac{D(r)H(r) + rD'(r)H(r) - rD(r)\left(\frac{d-1}{r}H(r) + 2D(r)\right)}{H^2(r)}$$

$$= \frac{-(d-2)D(r)H(r) + rD'(r)H(r) - 2rD^2(r)}{H^2(r)}$$

$$= \frac{2r}{H^2(r)}\left(H(r)\int_{\partial B_r} \left(\frac{\partial u}{\partial n}\right)^2 \, d\mathcal{H}^{d-1} - D^2(r)\right). \qquad (11.14)$$

Notice that, since u is harmonic in Ω_u, we have

$$D(r) = \int_{B_r} |\nabla u|^2 \, dx = \int_{\partial B_r} u\frac{\partial u}{\partial n} \, d\mathcal{H}^{d-1},$$

and so, by the Cauchy-Schwarz inequality and (11.14) we obtain $N'(r) \geq 0$.   □

Remark 11.8 (The Derivative of $\ln H$) Notice that, by (11.13), we get that

$$\frac{d}{dr}\left[\log\left(\frac{H(r)}{r^{d-1}}\right)\right] = 2\frac{N(r)}{r}. \qquad (11.15)$$

We are now in position to prove Proposition 11.4.

Proof of Proposition 11.4 Let $x_0 = 0 \in D$. We set $H(r)$, $D(r)$ and $N(r)$ to be as in Lemma 11.7 and Remark 11.8. Let $r_0 > 0$ be such that $B_{r_0}(x_0) \subset D$ and $H(r_0) > 0$. Since $u \in H^1(D)$, there is some $\varepsilon > 0$ such that $H > 0$ on the interval $(r_0 - \varepsilon, r_0)$. Then, the function $r \mapsto N(r)$ is non-decreasing in r and so

$$N(r) \leq N(r_0) \quad \text{for every} \quad r_0 - \varepsilon < r \leq r_0.$$

By (11.15), we have

$$\frac{d}{dr}\left[\log\left(\frac{H(r)}{r^{d-1}}\right)\right] = 2\frac{N(r)}{r} \le \frac{2N(r_0)}{r},$$ (11.16)

and integrating we get

$$\log\left(\frac{H(r_0)}{r_0^{d-1}}\right) - \log\left(\frac{H(r)}{r^{d-1}}\right) \le \log\left(\frac{r_0}{r}\right) 2N(r_0) \qquad \text{for every} \qquad r_0 - \varepsilon < r \le r_0.$$

This means that, for every $\varepsilon > 0$, H is bounded from below by a positive constant on the interval $[r_0 - \varepsilon, r_0]$. In particular, $H > 0$ on $(0, r_0]$. Thus, we can take $\varepsilon = r_0$.

Let now, $r \in (0, r_0/2]$. Integrating the inequality (11.16) from r to $2r$, we get

$$\log\left(\frac{H(2r)}{H(r)}\right) \le (d-1)\log 2 + 2\log 2\, N(r_0).$$

This implies that

$$H(2r) \le 2^{d-1}4^{N(r_0)} H(r) \qquad \text{for every} \qquad 0 < r \le \frac{r_0}{2}.$$

Integrating once more in r we get

$$\int_{B_{2r}} u^2\, dx \le 2^{d-1}4^{N(r_0)} \int_{B_r} u^2\, dx \qquad \text{for every} \qquad 0 < r \le \frac{r_0}{2}.$$ (11.17)

We next prove a Caccioppoli inequality for u in the ball B_{2r}. Indeed, let $\phi \in C_c^\infty(\mathbb{R}^d)$ be a cut-off function such that

$$\phi = 1 \quad \text{in} \quad B_r, \quad \phi = 0 \quad \text{on} \quad \mathbb{R}^d \setminus B_{2r}, \quad 0 \le \phi \le 1 \quad \text{and} \quad |\nabla\phi| \le 2/r \quad \text{in} \quad B_{2r} \setminus B_r.$$

Since, u is harmonic in Ω_u, the following Caccioppoli inequality does hold:

$$\int_{B_r} |\nabla u|^2\, dx \le \int_{B_{2r}} |\nabla(u\phi)|^2\, dx = \int_{B_{2r}} \left(u^2|\nabla\phi|^2 + \nabla u \cdot \nabla(u\phi^2)\right) dx$$

$$= \int_{B_{2r}} u^2|\nabla\phi|^2\, dx - \int_{B_{2r}} u\phi^2 \Delta u\, dx = \int_{B_{2r}} u^2|\nabla\phi|^2\, dx \le \frac{4}{r^2}\int_{B_{2r}} u^2\, dx.$$

On the other hand, by Lemma 11.6, there is a dimensional constant $C_d > 0$ such that:

$$\int_{B_r} u^2\, dx \le C_d\, r^2\left(\frac{|\Omega_u \cap B_r|}{|B_r|}\right)^{2/d} \int_{B_r} |\nabla u|^2\, dx \qquad \text{whenever} \qquad \frac{|\Omega_u \cap B_r|}{|B_r|} \le \frac{1}{2}.$$

This, together with the Caccioppoli and the doubling inequality (11.17), gives that

$$\int_{B_r} u^2 \, dx \le C_d \left(\frac{|\Omega_u \cap B_r|}{|B_r|} \right)^{2/d} 4^{N(r_0)} \int_{B_r} u^2 \, dx.$$

Since, $\displaystyle\int_{B_r} u^2 \, dx > 0$, we get that there is a dimensional constant C_d such that

$$\min \left\{ \frac{1}{2}, \frac{1}{C_d 2^{N(r_0)d}} \right\} \le \frac{|\Omega_u \cap B_r|}{|B_r|} \qquad \text{for every} \qquad 0 < r \le \frac{r_0}{2}.$$

In particular, we have a lower density bound for Ω_u at *every* point of D, which implies that $|D \setminus \Omega_u| = 0$ and concludes the proof. □

11.4   Convergence of the Lagrange Multipliers

In this section we prove that the Lagrange multipliers, associated to the solutions of variational problems with measure constraint in a fixed connected open set $D \subset \mathbb{R}^d$, are continuous with respect to variations of the constraint. This fact will be used several times in the proof of the optimality of the blow-up limits. In the next Lemma, which comes directly from [46], we will use the notation

$$\delta \text{Vol}\,(\Omega)[\xi] := \int_\Omega \text{div}\,\xi \, dx,$$

for every Lebesgue measurable set $\Omega \subset \mathbb{R}^d$ and every vector field $\xi \in C_c^\infty(\mathbb{R}^d; \mathbb{R}^d)$.

Lemma 11.9 (Convergence of the Lagrange Multipliers) *Let D be a connected bounded open set in $\mathbb{R}^d$ and let $u \in H_0^1(D)$ be a non-negative function for which (11.3) does hold. Let Λ_u be the Lagrange multiplier from (11.4) in D.*

Let $B \subset D$ be a connected bounded open set such that $0 < m := |\Omega_u \cap B| < |B|$. Let $(m_n)_{n \ge 1}$ be a sequence such that $\lim\limits_{n \to \infty} m_n = m$ and let $u_n \in H^1(B)$ be a solution (which exists due to Proposition 11.1) to the problem

$$\min \left\{ \mathcal{F}_0(v, B) \,:\, v \in H^1(B), \, v - u \in H_0^1(B), \, |\Omega_v| = m_n \right\}. \tag{11.18}$$

Then, we have:

(i) *for every n, there is a Lagrange multiplier* $\Lambda_{u_n} > 0$ *for which*

$$\delta\mathcal{F}_0(u_n, B)[\xi] + \Lambda_{u_n}\delta\mathrm{Vol}\,(\Omega_{u_n})[\xi] = 0 \qquad for\ every \qquad \xi \in C_c^\infty(B; \mathbb{R}^d),$$
$$(11.19)$$

(ii) *for every n, there is a vector field* $\xi_n \in C_c^\infty(B; \mathbb{R}^d)$ *such that*

$$\delta\mathcal{F}_0(u_n, B)[\xi_n] + \Lambda_{u_n} = 0 \qquad and \qquad \delta\mathrm{Vol}\,(\Omega_{u_n})[\xi_n] = 1. \qquad (11.20)$$

(iii) u_n *converges strongly in* $H_0^1(D)$ *and pointwise almost everywhere to a function* u_∞, *which is a solution to the problem*

$$\min\Big\{\mathcal{F}_0(v, B) \,:\, v \in H^1(B),\ v - u \in H_0^1(B),\ |\Omega_v| = m\Big\}; \qquad (11.21)$$

(iv) *the sequence of characteristic functions* $\mathbb{1}_{\Omega_{u_n}}$ *converges to* $\mathbb{1}_{\Omega_{u_\infty}}$ *pointwise almost everywhere and strongly in* $L^2(D)$;

(v) $\lim\limits_{n\to\infty} \Lambda_{u_n} = \Lambda_{u_\infty}$, *where* $\Lambda_{u_\infty} > 0$ *is the Lagrange multiplier of* u_∞ *in B, that is,*

$$\delta\mathcal{F}_0(u_\infty, B)[\xi] + \Lambda_{u_\infty}\delta\mathrm{Vol}\,(\Omega_{u_\infty})[\xi] = 0 \qquad for\ every \qquad \xi \in C_c^\infty(B; \mathbb{R}^d),$$
$$(11.22)$$

(vi) *Suppose that* $B \neq D$ *and that there is a connected component* C *of* $D \setminus \overline{B}$ *such that*

$$0 < |\Omega_u \cap C| < |C|,$$

then $\Lambda_{u_\infty} = \Lambda_u$.

Proof The existence of a solution u_n follows from Proposition 11.1. The existence of a Lagrange multiplier Λ_{u_n} and a vector field $\xi_n \in C_c^\infty(B; \mathbb{R}^d)$ with the properties (11.20) follows by Proposition 11.2. Let $\xi_0 \in C_c^\infty(B; \mathbb{R}^d)$ be a vector field such that

$$\delta\mathcal{F}_0(u, B)[\xi_0] + \Lambda_u = 0 \qquad and \qquad \delta\mathrm{Vol}\,(\Omega_u)[\xi_0] = 1.$$

Setting $u_t := u \circ (Id + t\xi_0)^{-1}$, we get that, for t small enough, $u_t - u \in H_0^1(D)$. Moreover, to every n large enough, we can associate a unique $t_n \in \mathbb{R}$ such that

$$u - u_{t_n} \in H_0^1(B) \qquad and \qquad |\Omega_{u_n}| = m_n = |\Omega_{u_{t_n}}|.$$

Thus, we can use u_{t_n} as a test function in (11.18). Thus, there is a constant C depending only on u and ξ_0 such that, for every n large enough (say $n \geq n_0$ for some $n_0 \in \mathbb{N}$), we have

$$\mathcal{F}_0(u_n, B) \leq \mathcal{F}_0(u_{t_n}, B) \leq C.$$

Then the sequence $(u_n)_{n \geq 1}$ is uniformly bounded in $H^1(B)$ and so, up to a subsequence, u_n converges weakly in H^1, strongly in L^2 and pointwise almost everywhere to a function $u_\infty \in H^1(B)$ such that $u_\infty - u \in H_0^1(B)$. In particular, the pointwise convergence of u_n to u_∞ implies that

$$\mathbb{1}_{\Omega_{u_\infty}} \leq \liminf \mathbb{1}_{\Omega_{u_n}}.$$

Thus, we get that

$$|\Omega_{u_\infty}| \leq \liminf m_n = m,$$

and so, the minimality of u implies that

$$\mathcal{F}_0(u, B) \leq \mathcal{F}_0(u_\infty, B).$$

On the other hand, the weak H^1 convergence of u_n gives that

$$\mathcal{F}_0(u_\infty, B) \leq \liminf_{n \to \infty} \mathcal{F}_0(u_n, B) \leq \liminf_{n \to \infty} \mathcal{F}_0(u_{t_n}, B) = \mathcal{F}_0(u, B),$$

so, we get $\mathcal{F}_0(u_\infty, B) = \mathcal{F}_0(u, B)$. Thus, u_∞ is a solution to (11.21) and $|\Omega_{u_\infty}| = m$. Moreover, using again the optimality of u_n and the fact that u_{t_n} converges to u, we obtain

$$\limsup_{n \to \infty} \mathcal{F}_0(u_n, B) \leq \limsup_{n \to \infty} \mathcal{F}_0(u_{t_n}, B) = \mathcal{F}_0(u, B) \leq \mathcal{F}_0(u_\infty, B) \leq \liminf_{n \to \infty} \mathcal{F}_0(u_n, B),$$

which gives that

$$\lim_{n \to \infty} \mathcal{F}_0(u_n, B) = \mathcal{F}_0(u_\infty, B).$$

Since u_n converges strongly in $L^2(B)$ and weakly in $H^1(B)$ to u_∞, we get that the convergence of u_n to u_∞ is strong in $H^1(B)$.

We next prove (iv). We will first show that $\mathbb{1}_{\Omega_{u_n}}$ convergence strongly in $L^2(B)$ to $\mathbb{1}_{\Omega_{u_\infty}}$.

Indeed, we first notice that, up to a subsequence, there is $h \in L^2(B)$ such that $\mathbb{1}_{\Omega_{u_n}} \rightharpoonup h$ weakly in $L^2(B)$. On the other hand, the pointwise convergence of u_n to u_∞ implies that

$$\mathbb{1}_{\Omega_{u_\infty}} \leq \liminf_{n \to \infty} \mathbb{1}_{\Omega_{u_n}}.$$

Thus, for any non-negative function $\varphi \in L^2(B)$, the Fatou Lemma implies that

$$\int_B \mathbb{1}_{\Omega_{u_\infty}} \varphi \, dx \leq \int_B \liminf \mathbb{1}_{\Omega_{u_n}} \varphi \, dx \leq \liminf \int_B \mathbb{1}_{\Omega_{u_n}} \varphi \, dx = \int_B h\varphi \, dx \,,$$

which yields $\mathbb{1}_{\Omega_{u_\infty}} \leq h$. In particular,

$$\|h\|^2_{L^2(B)} \geq |\Omega_{u_\infty}| = m.$$

On the other hand, the weak L^2 convergence of $\mathbb{1}_{\Omega_{u_n}}$ to h gives that

$$\|h\|^2_{L^2(B)} \leq \liminf_{n \to \infty} \|\mathbb{1}_{\Omega_{u_n}}\|^2_{L^2(B)} = \lim_{n \to \infty} m_n = m.$$

As a consequence,

$$\|h\|^2_{L^2(B)} = \lim_{n \to \infty} \|\mathbb{1}_{\Omega_{u_n}}\|^2_{L^2(B)} = m,$$

which implies that $\mathbb{1}_{\Omega_{u_n}}$ converges to h strongly in $L^2(B)$. Now, since

$$\mathbb{1}_{\Omega_{u_\infty}} \leq h \qquad \text{and} \qquad \|h\|^2_{L^2(B)} = |\Omega_{u_\infty}| = m,$$

we get that $h = \mathbb{1}_{\Omega_{u_\infty}}$, from which we conclude that $\mathbb{1}_{\Omega_{u_n}}$ converges to $\mathbb{1}_{\Omega_{u_\infty}}$ strongly in $L^2(B)$, and so, up to a subsequence $\mathbb{1}_{\Omega_{u_n}}$ converges to $\mathbb{1}_{\Omega_{u_\infty}}$ pointwise almost everywhere.

We now prove (v). We first notice that u and u_∞ are both solutions of (11.21). By Proposition 11.2, there is a Lagrange multiplier $\Lambda_\infty := \Lambda_{u_\infty} > 0$ such that (11.22) does hold. Moreover, by (iii) and (iv), we get that, for every fixed $\xi \in C_c^\infty(B; \mathbb{R}^d)$,

$$\delta \mathcal{F}_0(u_\infty, B)[\xi] = \lim_{n \to \infty} \delta \mathcal{F}_0(u_n, B)[\xi]$$

$$\delta \text{Vol}(\Omega_{u_\infty})[\xi] = \lim_{n \to \infty} \delta \text{Vol}(\Omega_{u_n})[\xi].$$

Now, choosing any $\xi \in C_c^\infty(B; \mathbb{R}^d)$ such that

$$\int_{\Omega_{u_\infty}} \text{div} \, \xi \, dx \neq 0,$$

and using (11.22) and (11.19) we get that Λ_{u_n} converges to Λ_∞.

We prove the last claim (vi). Indeed, since

$$\mathcal{F}_0(u, B) = \mathcal{F}_0(u_\infty, B) \qquad \text{and} \qquad |\Omega_{u_\infty}| = |\Omega_u \cap B| = m,$$

we get that the function

$$\widetilde{u} := \begin{cases} u_\infty & \text{in} \quad B \\ u & \text{in} \quad D \setminus B, \end{cases}$$

is in $H^1(D)$ and is a solution to the problem

$$\min \left\{ \mathcal{F}_0(v, D) \; : \; v \in H^1(D), \; v - u \in H_0^1(D), \; |\Omega_v| = |\Omega_u| \right\}.$$

In particular, $\widetilde{u}$ is a critical point of $\mathcal{F}_{\Lambda_\infty}$ in the entire D, that is,

$$\delta \mathcal{F}_0(\widetilde{u}, D)[\xi] + \Lambda_\infty \delta \text{Vol}\,(\Omega_{\widetilde{u}})[\xi] = 0 \qquad \text{for every} \qquad \xi \in C_c^\infty(D; \mathbb{R}^d).$$

On the other hand, in the connected component $\mathcal{C}$, we have that $\widetilde{u} = u$ and so, there is a vector field $\xi_0' \in C_c^\infty(\mathcal{C}; \mathbb{R}^d)$ such that $\delta \text{Vol}\,(\Omega_u)[\xi_0'] = \delta \text{Vol}\,(\Omega_{\widetilde{u}})[\xi_0'] = 1$. This implies that

$$\Lambda_\infty = \Lambda_\infty \delta \text{Vol}\,(\Omega_{\widetilde{u}})[\xi_0'] = -\delta \mathcal{F}_0(\widetilde{u}, D)[\xi_0']$$
$$= -\delta \mathcal{F}_0(u, D)[\xi_0'] = \Lambda_u \delta \text{Vol}\,(\Omega_u)[\xi_0'] = \Lambda_u \,,$$

which concludes the proof. $\square$

11.5   Almost Optimality of u at Small Scales

Let $D \subset \mathbb{R}^d$ be a connected bounded open set and $u : D \to \mathbb{R}$ be a non-negative function satisfying (11.3). In this section, we will prove the following result, which is analogous to the results of Briançon [5], Briançon-Lamboley [6], and the more recent [46], which are all dedicated to different (and technically more involved) free boundary problems arising in Shape Optimization.

Proposition 11.10 *Let D be a connected bounded open set in $\mathbb{R}^d$ and let $u \in H^1(D)$ be a non-negative function satisfying (11.3). Let $\Lambda > 0$ be the corresponding Lagrange multiplier, that is, Λ is such that $\delta \mathcal{F}_\Lambda(u, D) = 0$. Let $B \subset D$ be a ball such that:*

- $0 < |\Omega_u \cap B| < |B|$;
- $D \setminus \overline{B}$ *is connected:*
- $0 < |\Omega_u \cap D \setminus \overline{B}| < |D \setminus \overline{B}|.$

Then, for every $\varepsilon > 0$, there exists $r > 0$ such that u satisfies the following optimality conditions in every $B_r(x_0) \subset B$:

$$\mathcal{F}_{\Lambda+\varepsilon}(u, D) \leq \mathcal{F}_{\Lambda+\varepsilon}(v, D) \text{ for every } v \in H^1(D) \text{ such that } \begin{cases} v - u \in H_0^1(B_r(x_0)), \\ |\Omega_u| \leq |\Omega_v|. \end{cases}$$

$$(11.23)$$

$$\mathcal{F}_{\Lambda-\varepsilon}(u, D) \leq \mathcal{F}_{\Lambda-\varepsilon}(v, D) \text{ for every } v \in H^1(D) \text{ such that } \begin{cases} v - u \in H_0^1(B_r(x_0)), \\ |\Omega_u| \geq |\Omega_v|. \end{cases}$$

$$(11.24)$$

Remark 11.11 An immediate consequence of the inwards (11.24) and the outwards (11.23) optimality is that u satisfies the following almost-minimality condition:

$$\mathcal{F}_\Lambda(u, D) \leq \mathcal{F}_\Lambda(v, D) + \varepsilon|B_r| \text{ for every } v \in H^1(D)$$

$$\text{such that } v - u \in H_0^1(B_r(x_0)).$$

In order to prove Proposition 11.10 we will use the contradiction argument of Briançon [5]. The proof presented here follows step-by-step the exposition from [46] and uses only the existence of a Lagrange multiplier, variations with respect to smooth vector fields and elementary variational arguments. Roughly speaking, the main idea is to replace the localization condition $u - v \in H_0^1(B_r)$ in (11.24) and (11.23) by a bound on the measure of Ω_v, $|\Omega_v| \leq |\Omega_u| + \delta$, for which the passages to the limit are somehow easier. Proposition 11.10 is a direct consequence of Proposition 11.16

Remark 11.12 We notice that we work in the ball $B \subset D$ only because of the fact that we will use several times the convergence of the Lagrange multipliers associated to solutions of auxiliary problems. Indeed, in order to assure the convergence of these Lagrange multipliers to Λ (the Lagrange multiplier of the solution u), we need to work strictly inside the domain D (see Lemma 11.9, claim (vi)).

Definition 11.13 (Upper and Lower Lagrange Multipliers) We fix u, D and B to be as in Proposition 11.10. We set $m := |\Omega_u \cap B|$. For any constant $\delta > 0$, we define the upper Lagrange multiplier $\mu_+(\delta)$ as follows:

$$\mu_+(\delta) = \inf\{\mu \geq 0 \quad \text{for which (11.25) does hold}\}, \quad \text{where}$$

$$\mathcal{F}_\mu(u, B) \leq \mathcal{F}_\mu(v, B) \quad \text{for every} \quad v \in H^1(B) \quad \text{such that} \quad \begin{cases} u - v \in H_0^1(B), \\ m \leq |\Omega_v| \leq m + \delta. \end{cases}$$

$$(11.25)$$

Analogously, we define the lower Lagrange multiplier $\mu_-(\delta)$:

$$\mu_-(\delta) = \sup\left\{\mu \geq 0 \quad \text{for which (11.26) does hold}\right\}, \quad \text{where}$$

$$\mathcal{F}_\mu(u, B) \leq \mathcal{F}_\mu(v, B) \quad \text{for every} \quad v \in H^1(B) \quad \text{such that} \quad \begin{cases} u - v \in H_0^1(B), \\ m - \delta \leq |\Omega_v| \leq m. \end{cases}$$

$$(11.26)$$

Lemma 11.14 *Suppose that D is a connected bounded open set in $\mathbb{R}^d$ and that $u \in H^1(D)$ is a given non-negative function such that:*

(a) $u \neq 0$ and $|D \setminus \Omega_u| > 0$;
(b) u is stationary for $\mathcal{F}_\Lambda$, that is,

$$\delta\mathcal{F}_\Lambda(u, D)[\xi] = 0 \quad \text{for every} \quad \xi \in C_c^\infty(D; \mathbb{R}^d).$$

Then, we have the following claims:

(i) Suppose that there are $\delta > 0$ and $\mu > 0$ such that u satisfies (11.25). Then, $\Lambda \leq \mu$.
(ii) Suppose that there are $\delta > 0$ and $\mu > 0$ such that u satisfies (11.26). Then, $\Lambda \geq \mu$.

Proof Let us first prove claim (i). By Lemma 11.3 and the hypothesis (a), we get that there is a vector field $\xi \in C_c^\infty(D; \mathbb{R}^d)$ such that

$$\delta\mathrm{Vol}\,(\Omega_u)[\xi] = \frac{d}{dt}\Big|_{t=0}|\Omega_{u_t}| = 1,$$

where $u_t = u \circ (Id + t\xi)^{-1}$. Since for t small enough, we have that $|\Omega_u| \leq |\Omega_{u_t}| \leq |\Omega_u| + \delta$, the minimality of u gives that

$$\mathcal{F}_\mu(u, D) \leq \mathcal{F}_\mu(u_t, D).$$

Thus, taking the derivative at $t = 0$, we get that

$$0 \leq \frac{d}{dt}\Big|_{t=0}\mathcal{F}_\mu(u_t, D) = \frac{d}{dt}\Big|_{t=0}\mathcal{F}_\Lambda(u_t, D) + (\mu - \Lambda)\frac{d}{dt}\Big|_{t=0}|\Omega_{u_t}| = \mu - \Lambda,$$

which proves (i). The proof of (ii) is analogous. $\qquad\square$

As an immediate corollary, we obtain the following lemma.

Lemma 11.15 ($\mu_- \leq \Lambda_u \leq \mu_+$) *Let D be a connected bounded open set in $\mathbb{R}^d$ and $u \in H^1(D)$ be a non-negative function such that (11.3) holds. Let $m = |\Omega_u|$*

and $\Lambda_u > 0$ be the Lagrange multiplier of u in D, that is,

$$\delta\mathcal{F}_0(u, D)[\xi] + \Lambda_u \delta Vol(\Omega_u)[\xi] \qquad for\ every \qquad \xi \in C_c^\infty(D; \mathbb{R}^d).$$

Let B, $\mu_+(\delta)$ and $\mu_-(\delta)$ be as in Definition 11.13. Then, for every $\delta > 0$, we have

$$\mu_-(\delta) \le \Lambda_u \le \mu_+(\delta).$$

Notice that we still might have $\mu_-(\delta) = 0$ and $\mu_+(\delta) = +\infty$. In Proposition 11.16 below we will prove that this cannot occur.

Proposition 11.16 (Convergence of the Upper and the Lower Lagrange Multipliers) *Let D be a connected bounded open set in $\mathbb{R}^d$. Let $u \in H^1(D)$ be a non-negative function satisfying the minimality condition (11.3) in D and let $\Lambda_u > 0$ be the Lagrange multiplier of u in D, given by Proposition 11.2. Let $B \subset D$ be as in Proposition 11.10. Then, we have*

$$\lim_{\delta \to 0} \mu_+(\delta) = \lim_{\delta \to 0} \mu_-(\delta) = \Lambda_u.$$

Proof We will work only in the ball B. The presence of the larger domain D is only necessary to assure the convergence of the Lagrange multipliers (see Lemma 11.9) for the different auxiliary problems that we will use below. We will proceed in three steps.

Step 1 *We will first prove that $\mu_+(\delta) < +\infty$, for every $\delta > 0$.* This is equivalent to prove that there is some $\mu > 0$, for which the minimality condition (11.25) is satisfied. Assume, by contradiction, that for every $n > 0$, there exists some function $u_n \in H^1(B)$ such that

$$\mathcal{F}_n(u_n, B) < \mathcal{F}_n(u, B), \qquad u_n - u \in H_0^1(B) \qquad and \qquad m \le |\Omega_{u_n}| \le m + \delta.$$

Thus, if v_n is a solution of the auxiliary problem

$$\min\left\{\mathcal{F}_0(v, B) + n\left(|\Omega_v| - m\right)_+ : v \in H^1(B),\ v - u \in H_0^1(B),\ |\Omega_v| \le m + \delta\right\}, \tag{11.27}$$

then, we have that

$$\mathcal{F}_0(v_n, B) \le \mathcal{F}_0(v_n, B) + n\left(|\Omega_{v_n}| - m\right)_+ \le \mathcal{F}_n(u_n, B) + n\left(|\Omega_{u_n}| - m\right)_+ \tag{11.28}$$

$$< \mathcal{F}_0(u, B) + n\left(|\Omega_u| - m\right)_+ = \mathcal{F}_0(u, B).$$

Thus, by Proposition 11.1 (Step 5 of the proof), we obtain $|\Omega_{v_n}| > m$. Thus, we may assume

$$m < |\Omega_{v_n}| \le m + \delta \qquad for\ every \qquad n \in \mathbb{N}.$$

Using again (11.28), we obtain

$$\mathcal{F}_0(v_n, B) + n(|\Omega_{v_n}| - m) < \mathcal{F}_0(u, B),$$

which, in particular, implies that

$$|\Omega_{v_n}| - m \le \frac{1}{n}\mathcal{F}_0(u, B) \quad \text{and} \quad \lim_{n \to \infty} |\Omega_{v_n}| = m.$$

Now, notice that, setting $m_n := |\Omega_{v_n}|$, we have that v_n is a solution of

$$\min\left\{\mathcal{F}_0(v, B) \ : \ v \in H^1(B), \ v - u \in H_0^1(B), \ |\Omega_v| = m_n\right\}.$$

In particular, there is a Lagrange multiplier Λ_{v_n} such that

$$\delta\mathcal{F}_0(v_n, B)[\xi] + \Lambda_{v_n}\delta\mathrm{Vol}\,(\Omega_{v_n})[\xi] = 0 \quad \text{for every} \quad \xi \in C_c^\infty(B; \mathbb{R}^d),$$

and a vector field $\xi_n \in C_c^\infty(B; \mathbb{R}^d)$ such that

$$\delta\mathrm{Vol}\,(\Omega_{v_n})[\xi_n] = 1.$$

We set $v_n^t = v_n \circ (Id + t\xi_n)^{-1}$. Choosing $t > 0$ small enough and $n \in \mathbb{N}$ big enough, we get

$$v_n^t - u \in H_0^1(B) \quad \text{and} \quad m < |\Omega_{v_n^t}| < m + \frac{1}{n}\mathcal{F}_0(u, B) \le m + \delta.$$

Then, by (11.27), we have

$$\mathcal{F}_0(v_n, B) + n(|\Omega_{v_n}| - m) \le \mathcal{F}_0(v_n^t, B) + n(|\Omega_{v_n^t}| - m)$$
$$= \mathcal{F}_0(v_n, B) + t\,\delta\mathcal{F}_0(v_n, B)[\xi_n] + n(|\Omega_{v_n}| + t\,\delta\mathrm{Vol}\,(\Omega_{v_n})[\xi_n] - m) + o(t)$$
$$= \mathcal{F}_0(v_n, B) - t\Lambda_{v_n} + n(|\Omega_{v_n}| + t - m) + o(t),$$

which implies $n \le \Lambda_{v_n}$. On the other hand, Lemma 11.9 implies that

$$\lim_{n \to \infty} \Lambda_{v_n} = \Lambda_u < \infty,$$

which is a contradiction. This concludes the proof of Step 1.

Step 2 *In this step, we prove that* $\lim_{\delta \to 0} \mu_+(\delta) = \Lambda_u$.

Let δ_n be an infinitesimal decreasing sequence. We will prove that $\lim_{n \to \infty} \mu_+(\delta_n) = \Lambda_u$.

Fix $\varepsilon \in (0, \Lambda_u)$ and set α_n to be

$$0 < \alpha_n := \mu_+(\delta_n) - \varepsilon < \mu_+(\delta_n).$$

We will show that, for n big enough, $\alpha_n \leq \Lambda_u$. Let $u_n \in H^1(B)$ be solution to the auxiliary problem

$$\min\left\{ \mathcal{F}_0(v, B) + \alpha_n\big(|\Omega_v| - m\big)_+ \; : \; v \in H^1(B), \; v - u \in H_0^1(D), \; |\Omega_v| \leq m + \delta_n \right\}.$$
$$(11.29)$$

We consider two cases:

Case 1 (of Step 2). Suppose that $|\Omega_{u_n}| \leq m$. Then, the optimality of u gives

$$\mathcal{F}_0(u, B) \leq \mathcal{F}_0(u_n, B).$$

On the other hand, the optimality of u_n gives

$$\mathcal{F}_0(u, B) + \alpha_n\big(|\Omega_u| - m\big) = \mathcal{F}_0(u, B) + \alpha_n\big(|\Omega_u| - m\big)_+ \leq \mathcal{F}_0(u_n, B) + \alpha_n\big(|\Omega_{u_n}| - m\big)_+$$
$$\leq \mathcal{F}_0(v, B) + \alpha_n\big(|\Omega_v| - m\big)_+ = \mathcal{F}_0(v, B) + \alpha_n\big(|\Omega_v| - m\big),$$

for every $v \in H^1(B)$ such that $u - v \in H_0^1(B)$ and $m \leq |\Omega_v| \leq m + \delta_n$, which contradicts the definition of $\mu_+(\delta_n)$.

Case 2 (of Step 2). Suppose that $m < |\Omega_{u_n}| \leq m + \delta_n$. Notice that, setting $m_n := |\Omega_{u_n}|$, the solution u_n to (11.29) is also a solution to the problem

$$\min\left\{ \mathcal{F}_0(v, B) \; : \; v \in H^1(B), \; v - u \in H_0^1(B), \; |\Omega_v| = m_n \right\}.$$

By Proposition 11.2, there is a Lagrange multiplier $\Lambda_{u_n} \geq 0$ such that

$$\delta\mathcal{F}_0(u_n, B)[\xi] + \Lambda_{u_n}\delta\mathrm{Vol}\,(\Omega_{u_n})[\xi] = 0 \quad \text{for every} \quad \xi \in C_c^\infty(B_r; \mathbb{R}^d),$$

and a vector field $\xi_n \in C_c^\infty(B_r; \mathbb{R}^d)$ such that $\delta\mathrm{Vol}\,(\Omega_{u_n})[\xi_n] = 1$.

We set $u_n^t := u_n \circ (Id + t\xi_n)^{-1}$. By the minimality of u_n, for $t < 0$ small enough, we have

$$\mathcal{F}_0(u_n, B) + \alpha_n\big(|\Omega_{u_n}| - m\big) \leq \mathcal{F}_0(u_n^t, B) + \alpha_n\big(|\Omega_{u_n^t}| - m\big)$$
$$= \mathcal{F}_0(u_n, B) + t\,\delta\mathcal{F}_0(u_n, B)[\xi_n] + \alpha_n\big(|\Omega_{u_n}| + t\,\delta\mathrm{Vol}\,(\Omega_{u_n})[\xi_n] - m\big) + o(t)$$
$$= \mathcal{F}_0(u_n, B) - t\Lambda_{u_n} + \alpha_n\big(|\Omega_{u_n}| + t - m\big) + o(t),$$

from which we deduce that $\Lambda_{u_n} \geq \alpha_n$. Now, by Lemma 11.9 we get that

$$\lim_{n \to \infty} \mu_+(\delta_n) = \varepsilon + \lim_{n \to \infty} \alpha_n \leq \varepsilon + \lim_{n \to \infty} \Lambda_{u_n} = \varepsilon + \Lambda_u.$$

Since $\Lambda_u \leq \mu_+(\delta_n)$ and $\varepsilon > 0$ is arbitrary, we get the claim of Step 2.

Step 3 *In this last step we will prove that* $\lim_{\delta \to 0} \mu_-(\delta) = \Lambda_u$.
It is sufficient to show that, for aby decreasing infinitesimal sequence $\delta_n \to 0$, we have

$$\Lambda_u = \lim_{n \to \infty} \mu_-(\delta_n),$$

Precisely, we will show that for any fixed constant $\varepsilon > 0$, we have $\Lambda_u - \varepsilon \leq \lim_{n \to \infty} \mu_-(\delta_n)$.

Let $\beta_n := \mu_-(\delta_n) + \varepsilon$ and u_n be a solution of the problem

$$\min \left\{ \mathcal{F}_0(v, B) + \beta_n \big(|\Omega_v| - (m - \delta_n) \big)_+ \ : \ v \in H^1(B), \ v - u \in H_0^1(B), \ |\Omega_v| \leq m \right\}.$$

We consider three cases:
Case 1 (of Step 3). Suppose that $|\Omega_{u_n}| = m$.
By the minimality of u, we have that $\mathcal{F}_0(u, B) \leq \mathcal{F}_0(u_n, B)$. Now, the minimality of u_n, gives that, for every $v \in H^1(B)$ such that $v - u \in H_0^1(B)$ and $m - \delta_n \leq |\Omega_v| \leq m$, we have

$$\mathcal{F}_0(u, B) + \beta_n |\Omega_u| \leq \mathcal{F}_0(u_n, B) + \beta_n |\Omega_{u_n}| \leq \mathcal{F}_0(v, B) + \beta_n |\Omega_v|,$$

which contradicts the definition of $\mu_-(\delta_n)$.
Case 2 (of Step 3). Suppose that $|\Omega_{u_n}| < m - \delta_n$.
Then we have

$$\mathcal{F}_0(u_n, B) \leq \mathcal{F}_0(u_n + t\varphi, B),$$

for every $\varphi \in C_c^\infty(B)$ with sufficiently small compact support. This implies that u_n is harmonic in B. By the strong maximum principle, we obtain that either $u_n \equiv 0$ or $u_n > 0$ in B, which is impossible for n large enough.
Case 3 (of Step 3). Suppose that $m - \delta_n \leq |\Omega_{u_n}| < m$.
We set $m_n := |\Omega_{u_n}|$. Then, u_n is a solution to the problem

$$\min \left\{ \mathcal{F}_0(v, B) \ : \ v \in H^1(B), \ v - u \in H_0^1(B), \ |\Omega_v| = m_n \right\}.$$

By Proposition 11.2, there is a Lagrange multiplier $\Lambda_{u_n} \geq 0$ such that

$$\delta\mathcal{F}_0(u_n, B)[\xi] + \Lambda_{u_n}\delta\mathrm{Vol}\,(\Omega_{u_n})[\xi] = 0 \quad \text{for every} \quad \xi \in C_c^\infty(B; \mathbb{R}^d),$$

and a vector field $\xi_n \in C_c^\infty(B; \mathbb{R}^d)$ such that $\delta\mathrm{Vol}\,(\Omega_{u_n})[\xi_n] = 1$.
    We set $u_n^t := u_n \circ (Id + t\xi_n)^{-1}$. Let $t > 0$ be small enough. Then u_n^t is such that

$$u_n^t - v \in H_0^1(B) \qquad \text{and} \qquad |\Omega_{u_n}| = m_n \leq |\Omega_{u_n^t}| = m_n + t + o(t) < m\,.$$

Thus, by the minimality of u_n, we get

$$\begin{aligned}
\mathcal{F}_0(u_n, B) + \beta_n\big(|\Omega_{u_n}| - (m - \delta_n)\big) &\leq \mathcal{F}_0(u_n^t, B) + \beta_n\big(|\Omega_{u_n^t}| - (m - \delta_n)\big) \\
&\leq \mathcal{F}_0(u_n^t, B) + t\,\delta\mathcal{F}_0(u_n, B)[\xi_n] \\
&\quad + \beta_n\big(|\Omega_{u_n}| + t\,\delta\mathrm{Vol}\,(\Omega_{u_n})[\xi_n] - (m - \delta_n)\big) + o(t) \\
&= \mathcal{F}_0(u_n, B) - \Lambda_{u_n}t + \beta_n\big(|\Omega_{u_n}| + t - (m - \delta_n)\big) + o(t),
\end{aligned}$$

which implies that

$$\Lambda_{u_n} \leq \beta_n = \mu_-(\delta_n) + \varepsilon.$$

Finally, by Lemma 11.9, we get

$$\Lambda_u = \lim_{n\to\infty} \Lambda_{u_n} \leq \lim_{n\to\infty} \mu_-(\delta_n) + \varepsilon,$$

which concludes the proof. $\square$

Chapter 12
An Epiperimetric Inequality Approach to the Regularity of the One-Phase Free Boundaries

Throughout this chapter, we will use the notation

$$W_0(u) = \int_{B_1} |\nabla u|^2 \, dx - \int_{\partial B_1} u^2 \, d\mathcal{H}^{d-1} \quad \text{and} \quad W(u) = W_0(u) + |\{u > 0\} \cap B_1|,$$

where B_1 is the unit ball in $\mathbb{R}^d$, $d \geq 2$ and $u \in H^1(B_1)$.

The aim of this chapter is to prove an epiperimetric inequality for the energy W in dimension two. As a consequence, we will obtain the $C^{1,\alpha}$ regularity of the one-phase free boundaries in dimension two (see Proposition 12.13). Our main result is the following.

Theorem 12.1 (Epiperimetric Inequality for the Flat Free Boundaries) *There are constants $\delta_0 > 0$ and $\varepsilon > 0$ such that: if $c \in H^1(\partial B_1)$ is a non-negative function on the boundary of the disk $B_1 \subset \mathbb{R}^2$ and*

$$\pi - \delta_0 \leq \mathcal{H}^1\big(\{c > 0\} \cap \partial B_1\big) \leq \pi + \delta_0,$$

then, there exists a (non-negative) function $h \in H^1(B_1)$ such that $h = c$ on ∂B_1 and

$$W(h) - \frac{\pi}{2} \leq (1 - \varepsilon)\Big(W(z) - \frac{\pi}{2}\Big), \tag{12.1}$$

$z \in H^1(B_1)$ being the one-homogeneous extension of c in B_1, that is,

$$z(x) = |x| c\, (x/|x|).$$

Remark 12.2 On the figures in this section, we will use the following convention:

- ■ is the support $\Omega_h = \{h > 0\}$ of the competitor h;
- ╲╲ is the support $\Omega_z = \{z > 0\}$ of the one-homogeneous function z;
- ╱ is the boundary $\partial \Omega_h$;

© The Author(s) 2023
B. Velichkov, *Regularity of the One-phase Free Boundaries*,
Lecture Notes of the Unione Matematica Italiana 28,
https://doi.org/10.1007/978-3-031-13238-4_12

- $\diagup$ is the boundary $\partial \Omega_z$;
- $\diagup$ is the common boundary $\partial \Omega_h \cap \partial \Omega_z$.

In Theorem 12.1 the main assumption on the trace c is that the set $\Omega_c \subset \partial B_1$ is close to the half-sphere. In [49, Theorem 1] the epiperimetric inequality was proved under the different assumption that the trace is non-degenerate. In fact, the epiperimetric inequality (12.1) holds without any assumption on the trace $c : \partial B_1 \to \mathbb{R}$ or its free boundary $\partial \Omega_c \subset \partial B_1$. Indeed, in the Appendix, we will prove the following result, which covers both Theorem 12.1 and [49, Theorem 1].

Theorem 12.3 (Epiperimetric Inequality) *There is a constant $\varepsilon > 0$ such that: If $c \in H^1(\partial B_1)$ is a non-negative function on the boundary of the disk $B_1 \subset \mathbb{R}^2$ then, there exists a (non-negative) function $h \in H^1(B_1)$ such that (12.1) holds and $h = c$ on ∂B_1.*

Remark 12.4 (The Epiperimetric Inequality in Dimension $d \geq 2$) In higher dimension, the epiperimetric inequality for the one-phase energy is still an open problem. We expect that it will still be true under the assumption that the spherical set Ω_c is close to the half-sphere with respect to the Hausdorff distance. Indeed, it is an immediate consequence from the results in [29] that the epiperimetric inequality holds when the free boundary Ω_c is a $C^{2,\alpha}$ regular graph (in the sphere) over the equator.

We stress that in higher dimension the epiperimetric inequality can hold only under some additional assumption on the distance from the trace to the half-plane solution. Indeed, if this was not the case (and so, the epiperimetric inequality was true in dimension d without any assumption on the trace), then the singular set would be empty in any dimension. This is due to the following remark.

Remark 12.5 (The Epiperimetric Inequality Implies Regularity in Any Dimension) We claim that if u is a local minimizer of $\mathcal{F}_\Lambda$ in a neighborhood of x_0 and

$$W(u_{r,x_0}) - \frac{\omega_d}{2} \leq (1 - \varepsilon)\left(W(z_{r,x_0}) - \frac{\omega_d}{2} \right), \tag{12.2}$$

holds, for every $r > 0$, then x_0 is a regular point. This is due to the following facts:

- A point $x_0 \in \partial \Omega_u$ is regular, of and only if, the Lebesgue density of Ω_u at x_0 is precisely equal to $1/2$ (see Lemma 9.22).
- There are no points of Lebesgue density smaller than $1/2$ (Lemma 9.22).

- The function $r \mapsto W(u_{r,x_0})$ is non-decreasing and the limit

$$\lim_{r \to 0} W(u_{r,x_0})$$

is precisely the Lebesgue density of Ω_u at x_0 (see Lemma 9.20); in particular

$$W(u_{r,x_0}) - \omega_d/2 \geq 0 \qquad \text{for every} \qquad r \geq 0.$$

- Suppose that the epiperimetric inequality (12.2) holds for every $r > 0$. Then, by the Weiss formula (Lemma 9.2) we obtain the following bound on the energy

$$W(u_{r,x_0}) - \omega_d/2 \leq C r^\alpha,$$

for some $\alpha > 0$ depending on ε (this was proved in (12.28), which is the first step of the proof of Lemma 12.14). Since $W(u_{r,x_0}) - \frac{\omega_d}{2}$ is non-negative, we get that

$$\lim_{r \to 0} W(u_{r,x_0}) = \frac{\omega_d}{2}.$$

In particular, x_0 is a point of Lebesgue density $1/2$ and so, it should be a regular point, as mentioned in the first bullet above.

As a consequence of Remark 12.5 at the singular points of the free boundary (12.2) cannot hold, which means that in higher dimension the epiperimetric inequality can only be true under the additional assumption that the trace on ∂B_1 is close (in some sense) to a half-plane solution.

In this chapter, we will prove Theorem 12.1 and we will show that it implies the regularity of the free boundary (Proposition 12.13). The proof of Theorem 12.1 will be a consequence of the following two lemmas. The first one (Lemma 12.6) is based on a PDE argument which does not depend on the geometry of the free boundary; this lemma is proved in Sect. 12.5 and holds in any dimension $d \geq 2$. The second lemma (Lemma 12.7) reflects the interaction of the free boundary with the Dirichlet energy; we prove it in Sect. 12.3.3 and the proof strongly uses the fact that we work in dimension two, even if the main idea can be used also in dimension $d \geq 2$. Precisely, we use the Slicing Lemma (Lemma 12.10) to write the total energy as an integral of an energy defined on the spheres ∂B_r. Then, we prove the epiperimetric inequality by writing the second order expansion of the spherical energy for sets which are graphs over the equator (that is, arcs of length close to π).

Lemma 12.6 *Let ∂B_1 be the unit sphere in dimension $d \geq 2$. For every $\kappa > 0$, there are constants $\rho \in (0,1)$, $\varepsilon \in (0,1)$ and $\alpha > 1$, depending only on κ and d such that:*
If $\psi \in H^1(\partial B_1)$ satisfies the inequality

$$\int_{\partial B_1} |\nabla_\theta \psi|^2 \, d\mathcal{H}^{d-1} \geq (d - 1 + \kappa) \int_{\partial B_1} \psi^2 \, d\mathcal{H}^{d-1},$$

then, we have

$$W_0(h_\rho) \le (1 - \varepsilon) W_0(z) \qquad and \qquad W(h_\rho) \le (1 - \varepsilon) W(z), \qquad (12.3)$$

where in polar coordinates the functions z, $h_\rho : B_1 \to \mathbb{R}$ are given by

$$z(r, \theta) = r \psi(\theta) \qquad and \qquad h_\rho(r, \theta) = \left(\max\{r - \rho, 0\} \right)^\alpha \frac{\psi(\theta)}{(1 - \rho)^\alpha}.$$

Precisely, we can take

$$\varepsilon = \rho = \left(\frac{\kappa}{32 d^2 (2\kappa + 1)} \right)^3.$$

Lemma 12.7 (Epiperimetric Inequality for Principal Modes: The Flat Free Boundary Case) *Let B_1 be the unit ball in $\mathbb{R}^2$. There are constants $\delta_0 > 0$ and $\varepsilon > 0$ such that the following holds.*

If the continuous non-negative function $c : \partial B_1 \to \mathbb{R}$, $c \in H^1(\partial B_1)$, is a multiple of the first eigenfunction on $\{c > 0\} \subset \partial B_1$ and

$$\pi - \delta_0 \le \mathcal{H}^1(\{c > 0\} \cap \partial B_1) \le \pi + \delta_0,$$

then, there exists a (non-negative) function $h \in H^1(B_1)$ such that $h = c$ on ∂B_1 and

$$W(h) - \frac{\pi}{2} \le (1 - \varepsilon) \left(W(z) - \frac{\pi}{2} \right),$$

$z \in H^1(B_1)$ being the one-homogeneous extension of c in B_1. Moreover, if we assume that the function c is of the form

$$c(\theta) = c_1 \sin \left(\frac{\pi \theta}{\pi + \delta} \right) \mathbb{1}_{(0, \pi + \delta)}(\theta) \quad for \; some \quad c_1 > 0 \quad and \; some \quad \delta \in (-\delta_0, \delta_0),$$

then the one-homogeneous extension is given by $z(r, \theta) = r\, c(\theta)$ and the competitor h can be chosen as (the support of h is illustrated on Fig. 12.1)

$$h(r, \theta) = c_1 r \sin \left(\frac{\pi \theta}{\pi + t(r)} \right) \mathbb{1}_{(0, \pi + t(r))}(\theta), \quad where \quad t(r) = \left(1 - 3(1 - r)\varepsilon \right)\delta.$$

Fig. 12.1 The positivity sets Ω_h and Ω_z. Here, the trace c is a multiple of the first eigenfunction on the arc $(0, \pi + \delta)$, $|\delta| < \delta_0$ ($\delta < 0$ on the left and $\delta > 0$ on the right); the competitor is obtained by moving the free boundary $\partial\Omega_z$ towards the line $\{x_2 = 0\}$

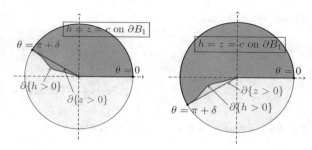

12.1 Preliminary Results

In this section we prove several preliminary results that we will use in the proof of Theorem 12.1 (and also in the proof of Theorem 12.3).

This section is organized as follows:

- In Lemmas 12.8 and 12.9 we discuss the scale-invariance and the decomposition of the energy in orthogonal directions; both these results are implicitly contained in [49].
- The Slicing Lemma (Lemma 12.10) shows how to disintegrate the energy along the different spheres ∂B_r, $0 < r < 1$. This result appeared for the first time in [29] and was crucial for the analysis of the free boundary around isolated singularities. We will use it in the proof of Lemma 12.7 (Sect. 12.3) and also in Sect. 12.2.

We start with the following result, which states that once we have a competitor for z in B_1, then we can rescale it and use it in any ball B_ρ ($\rho \le 1$) by attaching it to z at ∂B_ρ.

Lemma 12.8 (Scaling) *Suppose that* $z : B_1 \to \mathbb{R}$, $z(r, \theta) = rc(\theta)$ *is a one-homogeneous function and that* $h \in H^1(B_1)$ *is such that* $h = c = z$ *on* ∂B_1. *For every* $\rho \in (0, 1)$, *we set*

$$h_\rho(r, \theta) = \begin{cases} z(r, \theta) & \text{if } r \in [\rho, 1], \\ \rho\, h(r/\rho, \theta) & \text{if } r \in [0, \rho]. \end{cases}$$

then, we have

$$W(h_\rho) - W(z) = \rho^d\big(W(h) - W(z)\big).$$

Proof We first compute

$$W_0(h_\rho) - W_0(z) = \int_{B_1} |\nabla h_\rho|^2 \, dx - \int_{B_1} |\nabla z|^2 \, dx = \int_{B_\rho} |\nabla h_\rho|^2 \, dx - \int_{B_\rho} |\nabla z|^2 \, dx$$

$$= \rho^d \left(\int_{B_1} |\nabla h|^2 \, dx - \int_{B_1} |\nabla z|^2 \, dx \right) = \rho^d \left(W_0(h) - W_0(z) \right).$$

On the other hand, for the measure term, we have

$$|\{h_\rho > 0\} \cap B_1| - |\{z > 0\} \cap B_1| = |\{h_\rho > 0\} \cap B_\rho| - |\{z > 0\} \cap B_\rho|$$

$$= \rho^d \left(|\{h > 0\} \cap B_1| - |\{z > 0\} \cap B_1| \right),$$

which concludes the proof. □

Lemma 12.9 (Decomposition of the Energy) *Suppose that the functions* $h_1, h_2 \in H^1(B_1)$ *are such that, for every* $r \in (0, 1]$, *we have*

$$\int_{\mathbb{S}^{d-1}} \nabla_\theta h_1(r, \theta) \cdot \nabla_\theta h_2(r, \theta) \, d\theta = \int_{\mathbb{S}^{d-1}} h_1(r, \theta) h_2(r, \theta) \, d\theta = 0.$$

Then

$$W_0(h_1 + h_2) = W_0(h_1) + W_0(h_2).$$

Proof The claim follows directly from the definition of W_0 and the formula

$$\int_{B_1} |\nabla h|^2 \, dx = \int_0^1 r^{d-1} dr \int_{\partial B_1} \left(|\partial_r h|^2 + r^{-2} |\nabla_\theta h|^2 \right) d\theta,$$

which holds for any $h \in H^1(B_1)$. □

Lemma 12.10 (Slicing Lemma) *Let* B_1 *be the unit ball in* $\mathbb{R}^2$. *Let* $\phi : (0, 1] \times \mathbb{S}^1 \to \mathbb{R}$ *be a function such that* $\phi \in H^1((0, 1] \times \mathbb{S}^1)$. *Then, setting* $\phi(r, \theta) = \phi_r(\theta)$, *we have*

$$W_0(r\phi_r(\theta)) = \int_0^1 \mathcal{F}_0(\phi_r) \, r \, dr + \int_0^1 \int_{\mathbb{S}^1} \left(\partial_r \phi_r(\theta) \right)^2 r^3 dr,$$

and

$$W(r\phi_r(\theta)) = \int_0^1 \mathcal{F}(\phi_r) \, r \, dr + \int_0^1 \int_{\mathbb{S}^1} \left(\partial_r \phi_r(\theta) \right)^2 r^3 dr, \tag{12.4}$$

where, for any $\phi \in H^1(\mathbb{S}^1)$, we set

$$\mathcal{F}_0(\phi) = \int_{\mathbb{S}^1} \left(|\partial_\theta \phi|^2 - \phi^2 \right) d\mathcal{H}^1 \qquad and \qquad \mathcal{F}(\phi) = \mathcal{F}_0(\phi) + \mathcal{H}^1 \left(\{\phi > 0\} \cap \mathbb{S}^1 \right).$$

Proof Let $\phi :]0, 1] \times \partial B_1 \to \mathbb{R}$. Then,

$$W_0(r\phi_r(\theta)) = \int_0^1 \int_{\mathbb{S}^1} \left((\phi_r + r\partial_r \phi_r)^2 + (\partial_\theta \phi_r)^2 \right) d\theta \, r dr - \int_{\mathbb{S}^1} \phi_1^2(\theta) \, d\theta$$

$$= \int_0^1 \int_{\mathbb{S}^1} \left(\phi_r^2 + r\partial_r (\phi_r^2) + r^2 (\partial_r \phi_r)^2 + (\partial_\theta \phi_r)^2 \right) d\theta \, r dr - \int_{\mathbb{S}^1} \phi_1^2(\theta) \, d\theta$$

Integrating by parts, we get that

$$\int_0^1 r^2 \partial_r (\phi_r^2) \, dr = \phi_1^2 - 2 \int_0^1 \phi_r^2 \, r dr,$$

which implies that

$$W_0(r\phi_r(\theta)) = \int_0^1 \mathcal{F}_0(\phi_r) \, r dr + \int_0^1 \int_{\mathbb{S}^1} \left(\partial_r \phi_r(\theta) \right)^2 r^3 dr.$$

In order to prove (12.4), it is sufficient to notice that

$$|\{h > 0\} \cap B_1| = \int_0^1 \mathcal{H}^1 \left(\{\phi_r > 0\} \cap \mathbb{S}^1 \right) r dr,$$

where $h(r, \theta) = r\phi_r(\theta)$. $\qquad \square$

Remark 12.11 (The Energy of a One-Homogeneous Function) As an immediate consequence of Lemma 12.10, we get that if $c \in H^1(\partial B_1)$ and $z : B_1 \to \mathbb{R}$ is the one homogeneous extension of c in B_1, that is, $z(r, \theta) = rc(\theta)$, then

$$W_0(z) = \frac{1}{2} \mathcal{F}_0(c) \qquad and \qquad W(z) = \frac{1}{2} \mathcal{F}(c).$$

12.2   Homogeneity Improvement of the Higher Modes: Proof of Lemma 12.6

Let $\rho \in (0, 1)$ be fixed. We will first compute the energy of h_ρ. For this purpose, we will use the Slicing Lemma; for every $r \in [\rho, 1]$, we set

$$\phi_r(\theta) = \frac{\left(\max\{r - \rho, 0\}\right)^\alpha}{r} \frac{\psi(\theta)}{(1 - \rho)^\alpha}$$

and we compute

$$\mathcal{F}_0(\phi_r) = \frac{(r - \rho)^{2\alpha}}{r^2(1 - \rho)^{2\alpha}} \mathcal{F}_0(\psi) \quad \text{and}$$

$$\int_{\mathbb{S}^1} |\partial_r \phi_r|^2 \, d\theta = \left(\alpha - 1 + \frac{\rho}{r}\right)^2 \frac{(r - \rho)^{2\alpha - 2}}{r^2(1 - \rho)^{2\alpha}} \int_{\mathbb{S}^1} \psi^2 \, d\theta.$$

Integrating in r, we obtain

$$\int_\rho^1 \mathcal{F}_0(\phi_r) \, r^{d-1} dr = \frac{\mathcal{F}_0(\psi)}{(1 - \rho)^{2\alpha}} \int_\rho^1 (r - \rho)^{2\alpha} r^{d-3} dr$$

$$\leq \frac{\mathcal{F}_0(\psi)}{(1 - \rho)^{2\alpha}} \int_\rho^1 r^{2\alpha + d - 3} dr \leq \frac{1}{2\alpha + d - 2} \frac{\mathcal{F}_0(\psi)}{(1 - \rho)^{2\alpha}}. \tag{12.5}$$

We now compute

$$\int_\rho^1 \int_{\mathbb{S}^{d-1}} |\partial_r \phi_r|^2 \, d\theta \, r^{d+1} dr = \int_\rho^1 \left(\alpha - 1 + \frac{\rho}{r}\right)^2 \frac{(r - \rho)^{2\alpha - 2}}{r^2(1 - \rho)^{2\alpha}} r^{d+1} dr \int_{\mathbb{S}^{d-1}} \psi^2 \, d\theta$$

$$\leq \frac{2}{(1 - \rho)^{2\alpha}} \int_\rho^1 \left((\alpha - 1)^2 + \frac{\rho^2}{r^2}\right) r^{2\alpha + d - 3} dr \int_{\mathbb{S}^{d-1}} \psi^2 \, d\theta.$$

Integrating in $r \in [\rho, 1]$ and using that $\alpha \geq 1$ and $d \geq 2$, we get

$$\int_\rho^1 \left((\alpha - 1)^2 + \frac{\rho^2}{r^2}\right) r^{2\alpha + d - 3} dr \leq \frac{(\alpha - 1)^2}{2\alpha + d - 2} + \frac{\rho^2}{2\alpha + d - 4}$$

$$\leq \frac{1}{2} \left((\alpha - 1)^2 + \frac{\rho^2}{\alpha - 1}\right).$$

Together with the inequality

$$\int_{\mathbb{S}^{d-1}} \psi^2 \, d\theta \leq \frac{1}{\kappa} \mathcal{F}_0(\psi),$$

which we have by hypothesis, this implies

$$\int_{\rho}^{1} \int_{\mathbb{S}^{d-1}} |\partial_r \phi_r|^2 \, d\theta \, r^{d+1} dr \leq \frac{1}{(1-\rho)^{2\alpha}} \left((\alpha - 1)^2 + \frac{\rho^2}{\alpha - 1} \right) \frac{1}{\kappa} \mathcal{F}_0(\psi).$$

$$(12.6)$$

Furthermore, it is immediate to check that for every $\alpha \leq 2$ and $\rho \leq \frac{1}{2}$ we have

$$\frac{1}{(1-\rho)^{2\alpha}} \leq \frac{1}{(1-\rho)^4} \leq 1 + 128\rho \qquad \text{and} \qquad \frac{1}{(1-\rho)^{2\alpha}} \leq 16.$$

In particular,

$$\frac{(1-\rho)^{-2\alpha}}{2\alpha + d - 2} \leq \frac{1 + 128\rho}{2\alpha + d - 2} \leq \frac{1}{2\alpha + d - 2} + 64\rho,$$

which, together with (12.5) implies:

$$\int_{\rho}^{1} \mathcal{F}_0(\phi_r) \, r^{d-1} dr \leq \left(\frac{1}{2(\alpha - 1) + d} + 64\rho \right) \mathcal{F}_0(\psi). \qquad (12.7)$$

Analogously, from (12.6), we deduce

$$\int_{\rho}^{1} \int_{\mathbb{S}^{d-1}} |\partial_r \phi_r|^2 \, d\theta \, r^{d+1} dr \leq \frac{16}{\kappa} \left((\alpha - 1)^2 + \frac{\rho^2}{\alpha - 1} \right) \mathcal{F}_0(\psi). \qquad (12.8)$$

We are now in position to estimate the difference $W_0(h_\rho) - W_0(z)$. First of all, we set

$$\delta := \alpha - 1.$$

Using the identity (see Remark 12.11)

$$W_0(z) = \frac{1}{d} \mathcal{F}_0(\psi),$$

and the inequalities (12.7) and (12.8), we estimate

$$W_0(h_\rho) - W_0(z) \le \left(\frac{d}{2\delta + d} + 64d\,\rho + \frac{16d}{\kappa} \left(\delta^2 + \frac{\rho^2}{\delta} \right) - 1 \right) W_0(z)$$

$$\le \left(-\frac{2\delta}{d} + 64d\,\rho + \frac{16d}{\kappa} \left(\delta^2 + \frac{\rho^2}{\delta} \right) \right) W_0(z). \qquad (12.9)$$

We now choose

$$\rho = \delta^{3/2} \qquad \text{and} \qquad \delta^{1/2} = \frac{1}{32d^2(2 + 1/\kappa)}.$$

Substituting in (12.9), we obtain

$$W_0(h_\rho) - W_0(z) \le \left(-\frac{2\delta}{d} + 64d\delta^{3/2} + \frac{32d}{\kappa}\delta^2 \right) W_0(z)$$

$$\le \delta \left(-\frac{2}{d} + 32d(2 + 1/\kappa)\delta^{1/2} \right) W_0(z) \le -\frac{\delta}{d} W_0(z).$$

In particular, the first inequality in (12.3) holds for any $\varepsilon \le \delta/d$. In order to prove the second inequality in (12.3), we notice that, by the definition of h_ρ, we have

$$\left| \{h_\rho > 0\} \cap B_1 \right| = (1 - \rho) \left| \{z > 0\} \cap B_1 \right|.$$

Thus,

$$W(h_\rho) - W(z) = W_0(h_\rho) - W_0(z) + \left| \{h_\rho > 0\} \cap B_1 \right| - \left| \{z > 0\} \cap B_1 \right|$$

$$\le -\frac{\delta}{d} W_0(z) - \rho \left| \{z > 0\} \cap B_1 \right|.$$

Choosing

$$\varepsilon := \delta^{3/2},$$

we have that $\varepsilon \le \delta/d$ and so, we obtain

$$W(h_\rho) - W(z) \le -\varepsilon W(z),$$

which concludes the proof of (12.3).   □

12.3 Epiperimetric Inequality for the Principal Modes: Proof of Lemma 12.7

We suppose that the spherical set $\{c > 0\}$ is the arc $(0, \pi + \delta)$, where $\delta \in \mathbb{R}$ (and it might change sign). We recall that in Lemma 12.7 we assume that $|\delta| < \delta_0$. Then, we can write the trace c in the following form

$$c(\theta) = c_1 \phi_\delta(\theta) \quad \text{where} \quad c_1 > 0 \quad \text{and} \quad \phi_\delta(\theta) = \sin\left(\frac{\theta \pi}{\pi + \delta}\right) \quad \text{for} \quad \theta \in [0, \pi + \delta].$$

Next, for every $t \in \mathbb{R}$, we define the function $\phi_t : \mathbb{S}^1 \to \mathbb{R}$ as

$$\phi_t(\theta) = \sin\left(\frac{\theta \pi}{\pi + t}\right) \quad \text{for} \quad \theta \in [0, \pi + t], \qquad \phi_t(\theta) = 0 \quad \text{for} \quad \theta \notin [0, \pi + t].$$

Then set

$$f(t) := \int_{\partial B_1} \left(|\partial_\theta \phi_t(\theta)|^2 - \phi_t^2(\theta)\right) d\theta + \mathcal{H}^1(\{\phi_t > 0\}) - \pi$$

and

$$g(t) := \int_{\partial B_1} |\partial_t \phi_t(\theta)|^2 \, d\theta.$$

We consider the function

$$t(r) := \left(1 - 3(1 - r)\varepsilon\right)\delta,$$

and define the competitor h_δ as

$$h_\delta(r, \theta) = r\phi_{t(r)}(\theta), \tag{12.10}$$

which we will use in both Lemmas 12.7 and A.2.

We will show that for $\varepsilon > 0$ and $\delta > 0$ small enough, we have

$$W(c_1 h_\delta) - \frac{\pi}{2} \leq (1 - \varepsilon)\left(W(c_1 z_\delta) - \frac{\pi}{2}\right), \tag{12.11}$$

where z_δ is the one-homogeneous extension of ϕ_δ in B_1

$$z_\delta(r, \theta) = r\phi_\delta(\theta). \tag{12.12}$$

12.3.1 Reduction to the Case $c_1 = 1$

Let h_δ and z_δ be defined by (12.10) and (12.12). We claim that if, for some $\delta > 0$ and $\varepsilon > 0$, we have

$$W(h_\delta) - \frac{\pi}{2} \le (1 - \varepsilon)\left(W(z_\delta) - \frac{\pi}{2}\right), \qquad (12.13)$$

then (12.11) does hold for every $c_1 > 0$.

Indeed, using the homogeneity of W_0, we get that

$$W_0(c_1 z_\delta) = c_1^2 W_0(z_\delta) \qquad \text{and} \qquad W_0(c_1 h_\delta) = c_1^2 W_0(h_\delta).$$

On the other hand, we have that

$$\left|\{u > 0\} \cap B_1\right| = \int_0^1 \mathcal{H}^1\left(\{u > 0\} \cap \partial B_r\right) dr,$$

for every (continuous) function $u : B_1 \to \mathbb{R}$. Thus,

$$\left|\{z_\delta > 0\} \cap B_1\right| = \int_0^1 \mathcal{H}^1\left(\{\phi_\delta > 0\} \cap \partial B_r\right) dr = \int_0^1 \mathcal{H}^1\left(\{\phi_\delta > 0\} \cap \partial B_1\right) r\,dr$$

$$= \frac{1}{2}(\pi + \delta).$$

The analogous computation for the competitor h_δ gives

$$\left|\{h_\delta > 0\} \cap B_1\right| = \int_0^1 \mathcal{H}^1\left(\{\phi_{t(r)} > 0\} \cap \partial B_1\right) r\,dr = \int_0^1 (\pi + t(r)) r\,dr.$$

Putting together these computations, we obtain

$$\left(W(c_1 h_\delta) - \frac{\pi}{2}\right) - (1 - \varepsilon)\left(W(c_1 z_\delta) - \frac{\pi}{2}\right) = c_1^2\left[\left(W(h_\delta) - \frac{\pi}{2}\right) - (1 - \varepsilon)\left(W(z_\delta) - \frac{\pi}{2}\right)\right]$$

$$+ (1 - c_1^2)\left(\int_0^1 t(r) r\,dr - (1 - \varepsilon)\int_0^1 \delta r\,dr\right)$$

$$= c_1^2\left[\left(W(h_\delta) - \frac{\pi}{2}\right) - (1 - \varepsilon)\left(W(z_\delta) - \frac{\pi}{2}\right)\right] \le 0,$$

where we used that the function $t(r)$ is chosen in such a way that, for any δ and ε, we have

$$\int_0^1 \left(t(r) - (1 - \varepsilon)\delta\right) r\,dr = \delta \int_0^1 \left((1 - 3(1 - r)\varepsilon) - (1 - \varepsilon)\right) r\,dr$$

$$= \delta\varepsilon \int_0^1 \left(3r^2 - 2r\right) dr = 0.$$

The rest of Sect. 12.3 is dedicated to the proof of (12.13).

12.3.2 An Estimate on the Energy Gain

The Slicing Lemma (Lemma 12.10) implies that

$$W(h_\delta) = \int_0^1 f(t(r))\, r\, dr + \int_0^1 |t'(r)|^2 g(t(r))\, r^3\, dr \quad \text{and} \quad W(z_\delta) = \frac{1}{2} f(\delta).$$

We first notice that the error term $\int_0^1 |t'(r)|^2 g(t(r))\, r^3 dr$ is lower order. Precisely, we have

$$\int_0^1 |t'(r)|^2 g(t(r))\, r^3\, dr = 9\varepsilon^2\delta^2 \int_0^1 (1-r)^2 g(t(r))\, r^3\, dr \le C\varepsilon^2\delta^2,$$

where C is a universal numerical constant. Thus, we get

$$\left(W(h_\delta) - \frac{\pi}{2}\right) - (1-\varepsilon)\left(W(z_\delta) - \frac{\pi}{2}\right) \le F(\delta) + C\varepsilon^2\delta^2, \tag{12.14}$$

where we have set

$$F(t) := \int_0^1 \Big(f\big((1 - 3(1-r)\varepsilon)t\big) - (1-\varepsilon)f(t)\Big)r\, dr. \tag{12.15}$$

We will show that F is always negative in a neighborhoods of $t = 0$. First of all, we notice that the function f can be explicitly computed.

12.3.3 Computation of f

We now compute

$$\begin{aligned}
f(t) &= \int_0^{\pi+t} \left(\left(\frac{\pi}{\pi+t}\right)^2 \cos^2\left(\frac{\theta\pi}{\pi+t}\right) - \sin^2\left(\frac{\theta\pi}{\pi+t}\right)\right)d\theta + t \\
&= \frac{\pi+t}{\pi}\int_0^\pi \left(\left(\frac{\pi}{\pi+t}\right)^2 \cos^2\theta - \sin^2\theta\right)d\theta + t \\
&= \frac{\pi+t}{2}\left(\left(\frac{\pi}{\pi+t}\right)^2 - 1\right) + t \\
&= \pi\left(\frac{1+t/\pi}{2}\left(\left(\frac{1}{1+t/\pi}\right)^2 - 1\right) + t/\pi\right)
\end{aligned}$$

$$= \frac{\pi}{2}\left(\frac{1}{1+X} - 1 + X\right) = \frac{\pi}{2}\frac{X^2}{1+X} = \frac{\pi}{2}\left(X^2 - \frac{X^3}{1+X}\right),$$

where we set for simplicity $X = t/\pi$. In particular, this implies that

$$f(0) = f'(0) = 0 \quad \text{and} \quad f''(0) = \frac{1}{\pi}. \tag{12.16}$$

Moreover, we have that

$$\left|f(t) - \frac{t^2}{2\pi}\right| \leq \frac{|t|^3}{\pi^2} \quad \text{for every} \quad -1/2 \leq t \leq 1/2. \tag{12.17}$$

12.3.4   Conclusion of the Proof of Lemma 12.7

Notice that, by using (12.16) and taking the derivative under the sign of the integral, we get that

$$F(0) = F'(0) = 0.$$

Moreover, for the second derivative, we obtain

$$F''(0) = f''(0) \int_0^1 \left((1 - 3(1-r)\varepsilon)^2 - (1-\varepsilon)\right) r\, dr$$

$$= f''(0) \int_0^1 \left(-6(1-r)\varepsilon + 9(1-r)^2\varepsilon^2 + \varepsilon\right) r\, dr$$

$$= f''(0) \int_0^1 \left(-5r\varepsilon + 6r^2\varepsilon + 9(1-r)^2\varepsilon^2 r\right) dr = -C_\varepsilon f''(0),$$

where we have set

$$C_\varepsilon = \frac{\varepsilon}{2}\left(1 - \frac{3\varepsilon}{2}\right).$$

Thus, the second order Taylor expansion of F in zero is given by

$$F(0) + F'(0)t + \frac{1}{2}F''(0)t^2 = -\frac{C_\varepsilon}{2\pi}t^2.$$

We will next show that

$$\left|F(t) + \frac{C_\varepsilon}{2\pi}t^2\right| \leq |t|^3 \quad \text{for every} \quad -1/2 \leq t \leq 1/2. \tag{12.18}$$

Indeed, using (12.17) we can compute

$$\left| F(t) + \frac{C_\varepsilon}{2\pi} t^2 \right| \leq \left| \int_0^1 \left(f\big((1 - 3(1 - r)\varepsilon)t\big) - \frac{t^2}{2} f''(0)\big(1 - 3(1 - r)\varepsilon\big)^2 \right) r\, dr \right|$$

$$+ (1 - \varepsilon) \left| \int_0^1 \left(f(t) - \frac{t^2}{2} f''(0) \right) r\, dr \right|$$

$$\leq \frac{|t|^3}{\pi^2} \left(\int_0^1 \big(1 - 3(1 - r)\varepsilon\big)^3 r\, dr + (1 - \varepsilon) \int_0^1 r\, dr \right) \leq \frac{|t|^3}{\pi^2},$$

which gives (12.18).

Now, using (12.14), we estimate

$$\left(W(h_\delta) - \frac{\pi}{2} \right) - (1 - \varepsilon) \left(W(z_\delta) - \frac{\pi}{2} \right) \leq F(\delta) + C\varepsilon^2 \delta^2$$

$$\leq -\frac{1}{2\pi}\left(1 - \frac{3\varepsilon}{2}\right) \frac{\varepsilon}{2} \delta^2 + |\delta|^3 + C\varepsilon^2 \delta^2$$

$$\leq \left(-\frac{1}{2\pi}\left(1 - \frac{3\varepsilon}{2}\right) \frac{\varepsilon}{2} + \delta_0 + C\varepsilon^2 \right) \delta^2,$$

where C is the numerical constant from (12.14) and we recall that, by hypothesis, $|\delta| \leq \delta_0$.

We now choose ε and δ_0.

We set $\varepsilon = 16\pi\delta_0$. In particular, if $0 < \delta_0 \leq \dfrac{1}{48\pi}$, then $1 - \dfrac{3\varepsilon}{2} \geq \dfrac{1}{2}$, and so

$$-\frac{1}{2\pi}\left(1 - \frac{3\varepsilon}{2}\right) \frac{\varepsilon}{2} + \delta_0 + C\varepsilon^2 \leq -2\delta_0 + \delta_0 + C\varepsilon^2 \leq -\delta_0 + 256\pi^2 C \delta_0^2,$$

which is negative, whenever $\delta_0 \leq \dfrac{1}{256\pi^2 C}$. This means that in the end, choosing

$$\delta_0 = \min\left\{ \frac{1}{48\pi}, \frac{1}{256\pi^2 C} \right\} \qquad \text{and} \qquad \varepsilon = 16\pi\delta_0,$$

(12.13) holds for every δ such that $|\delta| \leq \delta_0$. This concludes the proof of Lemma 12.7.   □

12.4 Proof of Theorem 12.1

Since $c \in H^1(\partial B_1)$, we have that c is continuous and so, the set $\{c > 0\}$ is a countable union of disjoint intervals (arcs), that is,

$$\{c > 0\} = \bigcup_{j \geq 1} \mathcal{I}_j$$

where, by hypothesis, we have

$$\pi - \delta_0 \leq \sum_{j \geq 1} |\mathcal{I}_j| \leq \pi + \delta_0,$$

where $|\mathcal{I}_j| = \mathcal{H}^1(\mathcal{I}_j)$ denotes the length of the interval $\mathcal{I}_j$. Now, we consider two cases:

Case 1. There is one interval, say $\mathcal{I}_1$, of length $|\mathcal{I}_1| \geq \pi - \delta_0$. See Fig. 12.2.
Case 2. All the intervals are shorter than $\pi - \delta_0$, that is, $|\mathcal{I}_j| \leq \pi - \delta_0$, for every $j \geq 1$. See Fig. 12.3.

We first notice that if $\phi \in H_0^1(\mathcal{I}_j)$, then

$$\int_{\mathcal{I}_j} |\nabla_\theta \phi|^2 \, d\theta \geq \frac{\pi^2}{|\mathcal{I}_j|^2} \int_{\mathcal{I}_j} \phi^2 \, d\theta.$$

In particular, if $|\mathcal{I}_j| \leq \pi - \delta_0$, then

$$\int_{\mathcal{I}_j} |\nabla_\theta \phi|^2 \, d\theta \geq \left(1 + \frac{\delta_0}{\pi}\right) \int_{\mathcal{I}_j} \phi^2 \, d\theta.$$

Thus, if we are in *Case 2*, then the epiperimetric inequality is an immediate consequence of Lemma 12.6 with $\kappa = \delta_0/\pi$.

Fig. 12.2 The supports of the one homogeneous extension z (in red) and the competitor h (in blue); the trace c falls in Case 1; the length of $\mathcal{I}_1$ is smaller than π

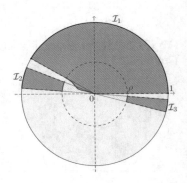

Fig. 12.3 The supports of the one homogeneous extension z (in red) and the competitor h (in blue) in Case 2

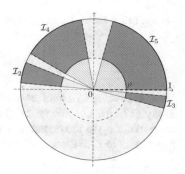

Suppose that we are in Case 1. Let $\{\phi_j\}_{j\geq 1}$ be a complete orthonormal system of eigenfunctions on the interval $\mathcal{I}_1$. For every $j \geq 1$, we set c_j to be the Fourier coefficient

$$c_j := \int_{\partial B_1} c(\theta)\phi_j(\theta)\,d\theta.$$

Then, we can decompose the trace c as

$$c(\theta) = c_1\phi_1(\theta) + \psi_1(\theta) + \psi_2(\theta),$$

where

$$\psi_1(\theta) = \sum_{j=2}^{\infty} c_j\phi_j(\theta),$$

and ψ_2 is the restriction of c on the set $\bigcup_{j\geq 2} \mathcal{I}_j$. We first claim that, for $i = 1, 2$, we have

$$\int_{\mathbb{S}^1} |\nabla_\theta \psi_i|^2\,d\theta \geq (1+\kappa)\int_{\mathbb{S}^1} \psi_i^2\,d\theta, \tag{12.19}$$

where $\kappa > 0$ is a constant depending only on δ_0. Indeed, since ψ_2 is supported on $\bigcup_{j\geq 2} \mathcal{I}_j$ and since $|\mathcal{I}_j| \leq 2\delta_0$, for $j \geq 2$, we have that

$$\int_{\mathbb{S}^1} |\nabla_\theta \psi_2|^2\,d\theta \geq \frac{\pi^2}{4\delta_0^2}\int_{\mathbb{S}^1} \psi_2^2\,d\theta. \tag{12.20}$$

On the other hand, ψ_1 contains only higher modes on the interval $\mathcal{I}_1$. Thus,

$$\int_{\mathcal{I}_1} |\nabla_\theta \psi_1|^2 \, d\theta \geq \frac{4\pi^2}{(\pi + \delta_0)^2} \int_{\mathcal{I}_1} \psi_1^2 \, d\theta. \tag{12.21}$$

Now, choosing δ_0 small enough (for instance, $\delta_0 \leq \pi/3$), (12.20) and (12.21) imply (12.19). Let now $\rho > 0$ and $\varepsilon_\psi > 0$ be the constants from Lemma 12.6 corresponding to the constant κ from (12.19). Let h_{ψ_1} and h_{ψ_2} be the competitors from Lemma 12.6 associated to the traces ψ_1 and ψ_2, respectively. Thus, we have

$$W_0(h_{\psi_1}) \leq (1 - \varepsilon_\psi) W_0(z_{\psi_1}) \qquad \text{and} \qquad W(h_{\psi_2}) \leq (1 - \varepsilon_\psi) W(z_{\psi_2}), \tag{12.22}$$

where $z_{\psi_i}(r, \theta) := z\psi_i(\theta)$.

Let $\tilde{h}$ be the competitor from Lemma 12.7, associated to the trace $c_1\phi_1$, and let

$$\tilde{z}(r, \theta) := rc_1\phi_1(\theta).$$

We set

$$\tilde{h}_\rho(r, \theta) = \begin{cases} \tilde{z}(r, \theta) & \text{if } r \in [\rho, 1], \\ \rho \tilde{h}(r/\rho, \theta) & \text{if } r \in [0, \rho]. \end{cases}$$

Thus, Lemmas 12.7 and 2.3 imply that

$$W(\tilde{h}_\rho) - \frac{\pi}{2} \leq (1 - \rho^d \tilde{\varepsilon}) \left(W(\tilde{z}) - \frac{\pi}{2} \right), \tag{12.23}$$

$\tilde{\varepsilon}$ being the constant from Lemma 12.7. We now define the competitor $h : B_1 \to \mathbb{R}$ as:

- $h = z$ if $W(z) \leq \pi/2$, where $z = \tilde{z} + z_{\psi_1} + z_{\psi_2}$ is the 1-homogeneous extension of c;
- $h = \tilde{z} + h_{\psi_1} + h_{\psi_2}$ if $W(z) > \pi/2$, but $W(\tilde{z}) \leq \pi/2$;
- $h = \tilde{h} + h_{\psi_1} + h_{\psi_2}$ if $W(z) > \pi/2$ and $W(\tilde{z}) > \pi/2$.

The first case is trivial and the second one follows directly by (12.22). We will prove the epiperimetric inequality in the most interesting third case. We first notice that the decomposition lemma (Lemma 12.9) implies that

$$W(z) = W(\tilde{z}) + W_0(z_{\psi_1}) + W(z_{\psi_1}),$$

and

$$W(h) = W(\tilde{h}_\rho) + W_0(h_{\psi_1}) + W(h_{\psi_1}),$$

where in the second decomposition, we use the fact that $h_{\psi_1} = h_{\psi_2} = 0$ in B_ρ and that $\tilde{h} = \tilde{z}$ outside B_ρ. Now, setting

$$\varepsilon = \min\{\rho^d \tilde{\varepsilon}, \varepsilon_\psi\},$$

the epiperimetric inequality (12.1) follows by (12.22) and (12.23).   □

12.5   Epiperimetric Inequality and Regularity of the Free Boundary

In this section we will show how the epiperimetric inequality (12.1) implies the $C^{1,\alpha}$ regularity of the free boundary. The main result of this section is Proposition 12.13, which we prove under the following assumption.

Condition 12.12 (Epiperimetric Inequality in Dimension $d \geq 2$) *We say that the epiperimetric inequality holds in dimension d if there are constants $\delta_d > 0$ and $\varepsilon_d > 0$ such that, for every non-negative one-homogeneous function $z \in H^1(B_1)$, which is δ_d-flat in the ball B_1 in some direction $\nu \in \partial B_1$, that is*

$$(x \cdot \nu - \delta_d)_+ \leq z(x) \leq (x \cdot \nu + \delta_d)_+ \quad \text{for every} \quad x \in B_1,$$

there exists a non-negative function $h : B_1 \to \mathbb{R}$ such that $z = h$ on ∂B_1 and

$$W(h) - \frac{\omega_d}{2} \leq (1 - \varepsilon_d)\left(W(z) - \frac{\omega_d}{2}\right). \tag{12.24}$$

Proposition 12.13 (ε-Regularity via Epiperimetric Inequality) *Suppose that the epiperimetric inequality holds in dimension d (that is, Condition 12.12 holds). Then, there is a constant $\varepsilon > 0$ such that if $u : B_1 \to \mathbb{R}$ is a non-negative minimizer of $\mathcal{F}_1$ in B_1 and is ε-flat in B_1 in some direction $\nu \in \partial B_1$*

$$(x \cdot \nu - \varepsilon)_+ \leq u(x) \leq (x \cdot \nu + \varepsilon)_+ \quad \text{for every} \quad x \in B_1,$$

then the free boundary $\partial \Omega_u$ is $C^{1,\alpha}$ regular in $B_{1/2}$.

Proof The claim is a consequence of Lemma 12.18, Lemma 12.14 and the results of the previous sections. By Condition 12.12 and Lemma 12.18, we have that the epiperimetric inequality (12.24) holds whenever

$$\|u_r - h_\nu\|_{L^2(B_2 \setminus B_1)}$$

is small enough for some half-plane solution h_ν.

Using this, together with the Weiss' monotonicity formula (Lemma 9.2), we get that the energy

$$\mathcal{E}(u) := W(u) - \frac{\omega_d}{2}$$

satisfies the hypotheses of Lemma 12.14. Thus, we obtain the uniqueness of the blow-up limit and the decay of the blow-up sequences at every point of the free boundary in $B_{1/2}$, that is, for every $x_0 \in B_{1/2}$, there is a function $u_{x_0} : \mathbb{R}^d \to \mathbb{R}$ such that

$$u_{x_0} = \lim_{r \to 0} u_{r,x_0} \quad \text{and} \quad \|u_{r,x_0} - u_{x_0}\|_{L^2(\partial B_1)} = 0.$$

Moreover, u_{x_0} is a global minimizer of $\mathcal{F}_1$ in $\mathbb{R}^d$ (Proposition 6.2) and is one-homogeneous (Proposition 9.12). Using again Lemma 12.14 (see the energy-decay estimate (12.28) in the first step of the proof), we get that

$$\lim_{r \to 0} \left(W(u_{r,x_0}) - \frac{\omega_d}{2} \right) = 0.$$

Thus, the strong convergence of the blow-up sequence u_{r,x_0} (Proposition 6.2) implies that

$$\frac{\omega_d}{2} = \lim_{r \to 0} W(u_{r,x_0}) = W(u_{x_0}).$$

By Lemma 9.22, we get that u_{x_0} is a half-plane solution. Thus, by Proposition 8.6, we get that the free boundary is a $C^{1,\alpha}$ regular in $B_{1/2}$. $\square$

The idea that a purely variational inequality as (12.1) encodes the local behavior of the free boundary goes back to Reifenberg [45] who proved the regularity of the area-minimizing surfaces via an epiperimetric inequality for the area functional. Weiss was the first to prove an epiperimetric inequality in the context of a free boundary problem; in [53] he proved such an inequality for the obstacle problem and recovered the $C^{1,\alpha}$ regularity of the (regular part of the) free boundary in any dimension, which was first proved by Caffarelli [11]. In [49], together with Luca Spolaor, we proved for the first time an epiperimetric inequality for the one-phase problem; in this case the interaction between the geometry of the free boundary and the Dirichlet energy functional is very strong and induced us to introduce the different *constructive* approach, which was the core of the previous section. In all these different contexts, once we have the epiperimetric inequality, we can obtain the regularity of the free boundary essentially by the same argument that we will describe in this section. The key result of this subsection is Lemma 12.14, which we attribute to Reifenberg, who was also the first to relate the variational epiperimetric inequality to the regularity of the local behavior of the free boundary (or area-minimizing surface).

Vocabulary and Notations We recall that, for any $u : B_1 \to \mathbb{R}$ and $r \leq 1$, we use the notation u_r to indicate the one-homogeneous rescaling of u

$$u_r(x) := \frac{1}{r} u(rx).$$

Then, if $\mathcal{E} : H^1(B_1) \to \mathbb{R}$ is a given energy (for instance, $\mathcal{E}(u) = W_1(u) - \frac{\omega_d}{2}$), we will use the following terminology:

- By **variation of the energy** we mean the variation, with respect to r, of the energy $\mathcal{E}$ of the rescaling u_r. In other words, the variation of the energy is simply

$$\frac{\partial}{\partial r} \mathcal{E}(u_r).$$

- The **energy deficit** of a function $v : B_1 \to \mathbb{R}$ is the difference

$$\mathcal{E}(v) - \mathcal{E}(u),$$

where $u : B_1 \to \mathbb{R}$ is a minimizer of $\mathcal{E}$ among all functions such that $u = v$ on ∂B_1.

- The **deviation** of a function $u : B_1 \to \mathbb{R}$ (from being one-homogeneous) is

$$\mathcal{D}(u) := \int_{\partial B_1} |x \cdot \nabla u(x) - u(x)|^2 \, d\mathcal{H}^{d-1}(x).$$

We notice that a function

$$u \in H^1(B_1) \quad \text{is one-homogeneous}$$

$$\Leftrightarrow \quad \mathcal{D}(u_r) = 0 \quad \text{for almost-every} \quad r \in (0, 1).$$

Lemma 12.14 (Reifenberg [45]) *Suppose that the function $u \in H^1(B_1)$ and the energy functional $\mathcal{E} : H^1(B_1) \to \mathbb{R}$ are such that:*

*(i) **Minimality.** $u_r \in H^1(B_1)$ minimizes $\mathcal{E}$ in B_1, for every $0 < r \leq 1$, that is,*

$$\mathcal{E}(u_r) \leq \mathcal{E}(v) \quad \text{for every} \quad v \in H^1(B_1), \quad v = u_r \quad \text{on} \quad \partial B_1.$$

*(ii) **The variation of the energy controls the deviation.** The function $r \mapsto \mathcal{E}(u_r)$ is non-negative, differentiable and there is a constant $C_2 > 0$ such that*

$$\frac{\partial}{\partial r} \mathcal{E}(u_r) \geq \frac{C_2}{r} \mathcal{D}(u_r) \quad \text{for every} \quad 0 < r < 1,$$

where $\mathcal{D}$ is given by

$$\mathcal{D}(u) := \int_{\partial B_1} |x \cdot \nabla u(x) - u(x)|^2 \, d\mathcal{H}^{d-1}(x).$$

(iii) **The variation of the energy controls the energy deficit of the homogeneous extension.** *There is a constant $C_3 > 0$ such that*

$$\frac{\partial}{\partial r}\mathcal{E}(u_r) \geq \frac{C_3}{r}\big(\mathcal{E}(z_r) - \mathcal{E}(u_r)\big) \qquad \text{for every} \qquad 0 < r < 1,$$

where $z_r : B_1 \to \mathbb{R}$ is the one-homogeneous extension of the trace $u_r|_{\partial B_1}$, that is,

$$z_r(x) = |x|u_r(x/|x|).$$

(iv) **Epiperimetric inequality.** *There is a one-homogeneous function $b : \mathbb{R}^d \to \mathbb{R}$ such that, if u_r is close to b in $B_2 \setminus B_1$, then an epiperimetric inequality holds in B_1. Precisely, there are constants $\varepsilon > 0$ and $\delta_4 > 0$ such that:*
For every $r \in (0, 1/2]$ satisfying

$$\|u_r - b\|_{L^2(B_2 \setminus B_1)} \leq \delta_4, \tag{12.25}$$

there is a function $h_r \in H^1(B_1)$ such that $h_r = u_r = z_r$ on ∂B_1 and

$$\mathcal{E}(h_r) \leq \big(1 - \varepsilon\big)\mathcal{E}(z_r). \tag{12.26}$$

Under the hypotheses (i), (ii), (iii) *and* (iv), *there is $\delta > 0$ such that, if u satisfies*

$$\sqrt{\mathcal{E}(u_1)} + \|u_1 - b\|_{L^2(B_1 \setminus B_{1/8})} \leq \delta$$

then there is a unique $u_0 \in H^1(B_1)$ such that

$$\|u_r - u_0\|_{L^2(\partial B_1)} \leq Cr^\gamma \qquad \text{for every} \qquad 0 < r \leq 1/2, \tag{12.27}$$

where the constants γ and C can be chosen as

$$\gamma = \frac{1}{2}\varepsilon C_3 \qquad \text{and} \qquad C = \delta_4.$$

Remark 12.15 If the epiperimetric inequality (12.26) holds without the closeness assumption (12.25), then the Step 4 of the proof of Lemma 12.14 can be omitted.

Remark 12.16 The energy to which we will apply Lemma 12.14 is the Weiss' boundary adjusted energy

$$\mathcal{E}(u) = W_1(u) = \int_{B_1} |\nabla u|^2 \, dx - \int_{\partial B_1} u^2 \, d\mathcal{H}^{d-1} + |\{u > 0\} \cap B_1|.$$

In this case, both *(ii)* and *(iii)* are implied by the Weiss' formula (Lemma 9.2).

Remark 12.17 In our case, the function b from assumption *(iv)* is the half-plane solution $b(x) = (x \cdot \nu)_+$ for some $\nu \in \partial B_1$. Notice, that this does not mean that the blow-up limit u_0 of u_r is equal to b. In fact, it may happen that the blow-up limit is another half-plane solution $b(x) = (x \cdot \tilde{\nu})_+$, with $\tilde{\nu}$, which is close to ν. More generally, this lemma can be applied to situations in which u_0 is not just a rotation of b, but is a completely different function. This happens for instance at isolated singularities in higher dimension (see [29]).

Proof of Lemma 12.14 Let now $0 \leq \rho \leq 1/2$ be the smallest non-negative number such that

$$\|u_r - b\|_{L^2(B_2 \setminus B_1)} < \delta_4 \qquad \text{for every} \qquad r \in (\rho, 1/2],$$

and so, we can apply the epiperimetric inequality (12.26) for every u_r with $r \in (\rho, 1/2]$. Notice that, since b is 1-homogeneous, a simple change of variables gives that

$$\|u_r - b\|^2_{L^2(B_2 \setminus B_1)} = r^{-(d+2)} \|u - b\|^2_{L^2(B_{2r} \setminus B_r)}.$$

Thus, by choosing $\delta < 4^{d+2}\delta_4$, we get that

$$\|u_r - b\|_{L^2(B_2 \setminus B_1)} = r^{-\frac{d+2}{2}} \|u - b\|_{L^2(B_{2r} \setminus B_r)} \leq 4^{d+2} \|u - b\|_{L^2(B_1 \setminus B_{1/8})},$$

for every $1/8 \leq r \leq 1/2$. Thus, $\rho \leq 1/8$.

We divide the proof in several steps.

Step 1: The Epiperimetric Inequality Implies the Decay of the Energy
Let $r \in (\rho, 1/2]$. By (iii), (iv) and the minimality of u_r (assumption (i)), we have

$$\frac{\partial}{\partial r} \mathcal{E}(u_r) \geq \frac{C_3}{r} \big(\mathcal{E}(z_r) - \mathcal{E}(u_r) \big)$$

$$\geq \frac{C_3}{r} \big(\mathcal{E}(h_r) + \varepsilon \, \mathcal{E}(z_r) - \mathcal{E}(u_r) \big) \geq \frac{\varepsilon C_3}{r} \mathcal{E}(u_r).$$

Setting $\gamma = \frac{1}{2}\varepsilon C_3$, we get that

$$\frac{\partial}{\partial r}\left(\frac{\mathcal{E}(u_r)}{r^{2\gamma}}\right) \geq 0,$$

and so,

$$\mathcal{E}(u_r) \leq 4^\gamma \mathcal{E}(u_{1/2})\, r^{2\gamma} \qquad \text{for every} \qquad r \in (\rho, 1/2]. \tag{12.28}$$

Step 2: The Energy Controls the Deviation We set

$$e(r) = \mathcal{E}(u_r) \qquad \text{and} \qquad f(r) = \mathcal{D}(u_r).$$

By (ii), we get that

$$\frac{C_2}{r^{1+\gamma}}f(r) \leq \frac{e'(r)}{r^\gamma} = \frac{\partial}{\partial r}\left(\frac{e(r)}{r^\gamma}\right) + \gamma\frac{e(r)}{r^{1+\gamma}} \leq \frac{\partial}{\partial r}\left(\frac{e(r)}{r^\gamma}\right) + \gamma\, 4^\gamma e(1/2)\, r^{2\gamma-1-\gamma},$$

which implies that for every $\rho < r_1 < r_2 \leq 1/2$, we have the estimate

$$C_2 \int_{r_1}^{r_2} \frac{f(r)}{r^{1+\gamma}}\, dr \leq \frac{e(r_2)}{r_2^\gamma} - \frac{e(r_1)}{r_1^\gamma} + 4^\gamma e(1/2)(r_2^\gamma - r_1^\gamma)$$

$$\leq \frac{e(r_2)}{r_2^\gamma} + 4^\gamma e(1/2)\, r_2^\gamma \leq 2\, 4^\gamma e(1/2)\, r_2^\gamma.$$

Step 3: The Deviation Controls the Oscillation of the Blow-up Sequence u_r
Let $x \in \partial B_1$ be fixed. Then, we have

$$\frac{\partial}{\partial r}(u_r(x)) = \frac{\partial}{\partial r}\left(\frac{u(rx)}{r}\right) = \frac{x \cdot \nabla u(rx)}{r} - \frac{u(rx)}{r^2} = \frac{1}{r}(x \cdot \nabla u_r(x) - u_r(x)).$$

Integrating in r, we get that, for every $0 < r_1 < r_2 \leq 1$,

$$|u_{r_2}(x) - u_{r_1}(x)| \leq \int_{r_1}^{r_2} \frac{1}{r}|x \cdot \nabla u_r(x) - u_r(x)|\, dr.$$

Integrating in $x \in \partial B_1$, and taking $\rho < r_1 < r_2 \leq 1/2$, we obtain

$$\int_{\partial B_1} |u_{r_2} - u_{r_1}|^2\, d\mathcal{H}^{d-1} \leq \int_{\partial B_1}\left(\int_{r_1}^{r_2} \frac{1}{r}|x \cdot \nabla u_r - u_r|\, dr\right)^2 d\mathcal{H}^{d-1}$$

$$\leq \int_{\partial B_1}\left(\int_{r_1}^{r_2} r^{\gamma-1}\, dt\right)\left(\int_{r_1}^{r_2} r^{-1-\gamma}|x \cdot \nabla u_r - u_r|^2\, dr\right) d\mathcal{H}^{d-1}$$

$$= \frac{1}{\gamma} \left(r_2^\gamma - r_1^\gamma \right) \left(\int_{r_1}^{r_2} r^{-1-\gamma} f(r) \, dr \right)$$

$$\leq \frac{1}{\gamma} r_2^\gamma \frac{1}{C_2} 2 \, 4^\gamma e(1/2) \, r_2^\gamma \leq \frac{8 \, e(1/2)}{\gamma C_2} r_2^{2\gamma}. \tag{12.29}$$

Step 4: The Blow-up Sequence Remains Close to b
Taking $r_2 \in (1/4, 1/2)$ and $r_1 = r \in (\rho, r_2)$ in (12.29), we get

$$\|u_r - b\|_{L^2(\partial B_1)} \leq \|u_r - u_{r_2}\|_{L^2(\partial B_1)} + \|u_{r_2} - b\|_{L^2(\partial B_1)} \leq \sqrt{\frac{8 \, e(1/2)}{\gamma C_2}} + \|u_{r_2} - b\|_{L^2(\partial B_1)}.$$

Now, since

$$\int_{B_{1/2} \setminus B_{1/4}} |u - b|^2 \, dx = \int_{1/4}^{1/2} \int_{\partial B_t} |u - b|^2 \, d\mathcal{H}^{d-1} \, dt$$

$$= \int_{1/4}^{1/2} t^{d+1} \|u_t - b\|_{L^2(\partial B_1)}^2 \, dt \geq 4^{-(d+1)} \int_{1/4}^{1/2} \|u_t - b\|_{L^2(\partial B_1)}^2 \, dt,$$

we can choose $r_2 \in (1/4, 1/2)$ such that

$$4^{d+1} \int_{B_{1/2} \setminus B_{1/4}} |u - b|^2 \, dx \geq \|u_{r_2} - b\|_{L^2(\partial B_1)}^2.$$

On the other hand, taking $r_1 = r \in (\rho, r_2/2)$, we obtain

$$\int_{B_2 \setminus B_1} |u_r - b|^2 \, dx = \int_1^2 \int_{\partial B_t} |u_r - b|^2 \, d\mathcal{H}^{d-1} \, dt = \int_1^2 t^{d+1} \|u_{rt} - b\|_{L^2(\partial B_1)}^2 \, dt$$

$$\leq 2^{d+1} \int_1^2 \|u_{rt} - b\|_{L^2(\partial B_1)}^2 \, dt$$

$$\leq 2^{d+2} \left(\sqrt{\frac{8 \, e(1/2)}{\gamma C_2}} + \|u_{r_2} - b\|_{L^2(\partial B_1)} \right)^2$$

$$\leq 2^{d+2} \left(\sqrt{\frac{8 \, e(1/2)}{\gamma C_2}} + 2^{d+1} \|u - b\|_{L^2(B_{1/2} \setminus B_{1/4})} \right)^2$$

This implies that if $u = u_1$ is such that

$$2^{d+2}\left(\sqrt{\frac{8\,e(1/2)}{\gamma\,C_2}} + 2^{d+1}\|u - b\|_{L^2(B_{1/2}\setminus B_{1/4})}\right) < \delta_4,$$

then $\rho = 0$, that is, the epiperimetric inequality (12.26) can be applied to every $r \in (0, 1]$.

Step 5: Conclusion As a consequence of the previous step, the decay estimate (12.29) holds on the whole interval $(0, 1]$:

$$\|u_{r_2} - u_{r_1}\|_{L^2(\partial B_1)} \le \sqrt{\frac{8\,e(1)}{\gamma\,C_2}}\,r_2^\gamma \qquad \text{for every} \qquad 0 < r_1 < r_2 \le 1/2. \qquad (12.30)$$

Thus, there is $u_0 \in L^2(\partial B_1)$, which is the strong $L^2(\partial B_1)$-limit of the blow-up sequence u_r

$$\lim_{r \to 0} u_r = u_0.$$

Finally, taking $r_2 = r \in (0, 1)$ and passing to the limit as $r_1 \to 0$ in (12.30), we obtain (12.27). This concludes the proof.

$\square$

In order to prove Proposition 12.13 under the Condition 12.12 we will need the following lemma.

Lemma 12.18 *For every $\varepsilon > 0$ there is $\delta > 0$ such that the following holds.*
If $u : B_2 \to \mathbb{R}$ is a (non-negative) minimizer of $\mathcal{F}_1$ in B_2 satisfying

$$\|u - h_\nu\|_{L^2(B_2\setminus B_1)} \le \delta \qquad \text{for some} \qquad \nu \in \partial B_1,$$

where h_ν is the half-plane solution $h_\nu(x) = (x \cdot \nu)_+$,
then u is ε-flat in the direction ν in the ball B_1, that is,

$$(x \cdot \nu - \varepsilon)_+ \le u(x) \le (x \cdot \nu + \varepsilon)_+ \qquad \text{for every} \qquad x \in B_1. \qquad (12.31)$$

Proof We will first prove that there is $\varepsilon > 0$ such that u is ε-flat on $\partial B_{3/2}$, that is,

$$(x \cdot \nu - \varepsilon)_+ \le u(x) \le (x \cdot \nu + \varepsilon)_+ \qquad \text{for every} \qquad x \in \partial B_{3/2}. \qquad (12.32)$$

From this, we will deduce that u is ε-flat in B_1.

In order to prove (12.32), we start by noticing that that, since u minimizes $\mathcal{F}_1$ in B_2, it is L-Lipschitz continuous in $B_{7/4}$, for some $L \ge 1$ depending only on the dimension (see Theorem 3.1). Then, also the function

$$u_\rho - h_\nu : B_{7/4} \to \mathbb{R}$$

is $(2L)$-Lipschitz continuous. Thus, there is a dimensional constant C_d such that

$$\|u_\rho - h_\nu\|_{L^\infty(B_{7/4} \setminus B_{5/4})} \le C_d L^{\frac{d}{d+2}} \|u_\rho - h_\nu\|_{L^2(B_{7/4} \setminus B_{5/4})}^{\frac{2}{d+2}} \le C_d \, \delta^{\frac{2}{d+2}}.$$

We now choose $\delta > 0$ such that

$$C_d \, \delta^{\frac{2}{d+2}} \le \varepsilon/2. \tag{12.33}$$

Thus,

$$\|u_\rho - h_\nu\|_{L^\infty(B_{7/4} \setminus B_{5/4})} \le \varepsilon/2. \tag{12.34}$$

Now, using (12.34), we obtain the estimate from below

$$(x \cdot \nu - \varepsilon)_+ \le u(x) \quad \text{for every} \quad x \in B_{7/4} \setminus B_{5/4},$$

while from above we only have

$$u(x) \le (x \cdot \nu + \varepsilon)_+ \quad \text{for every} \quad x \in \{x \cdot \nu \ge -\varepsilon/2\} \cap \left(B_{7/4} \setminus B_{5/4} \right).$$

Indeed, if $x \cdot \nu \ge -\varepsilon/2$, then

$$u(x) \le \varepsilon/2 + h_\nu(x) = \varepsilon/2 + (x \cdot \nu)_+ \le \begin{cases} (x \cdot \nu + \varepsilon)_+ & \text{if} \quad x \cdot \nu \ge 0, \\ \varepsilon/2 \le (x \cdot \nu + \varepsilon)_+ & \text{if} \quad 0 \ge x \cdot \nu \ge -\varepsilon/2. \end{cases}$$

Thus, in order to prove that (12.32) it is sufficient to show that

$$u = 0 \quad \text{on the set} \quad \{x \cdot \nu < -\varepsilon/2\} \cap \partial B_{3/2}. \tag{12.35}$$

On the other hand, u is also non-degenerate in the annulus $A := B_{7/4} \setminus B_{5/4}$, that is, there is a dimensional constant $0 < \kappa < 1$ such that (see Proposition 4.1)

$$x_0 \in A \cap \overline{\Omega}_u \quad \Rightarrow \quad \|u\|_{L^\infty(B_r(x_0))} \ge \kappa r \quad \text{for every} \quad r \le 1/4.$$

Suppose by absurd that there is a point

$$x_0 \in \overline{\Omega}_u \cap \{x \cdot \nu < -\varepsilon/2\} \cap \partial B_{3/2}.$$

Then, taking $r = \varepsilon/2$, we get that there is

$$y_0 \in B_r(x_0) \subset \{x \cdot v < 0\} \cap B_{7/4} \setminus B_{5/4}$$

such that

$$\left| u(y_0) - h_v(x_0) \right| = u(y_0) \geq \frac{1}{2} \kappa \varepsilon.$$

If we choose δ such that

$$C_d \, \delta^{\frac{2}{d+2}} \leq \frac{1}{2} \kappa \varepsilon, \tag{12.36}$$

then we reach a contradiction. Notice that, since $\kappa < 1$, (12.36) implies (12.33).

This concludes the proof of (12.32). The conclusion now follows by Proposition 12.19. $\qquad\square$

12.6   Comparison with Half-Plane Solutions

In this subsection, we prove the following result, which we use in the proof of Lemma 12.18; but is also of general interest.

Proposition 12.19 *Let $D \subset \mathbb{R}^d$ be a bounded open set and let $u : \overline{D} \to \mathbb{R}$ be a non-negative continuous function and a minimizer of the functional $\mathcal{F}_\Lambda$ in D. Let $c \in \mathbb{R}$ be a constant, $v \in \mathbb{R}^d$ be a unit vector and*

$$h(x) = \sqrt{\Lambda} \left(x \cdot v + c \right)_+$$

be a half-plane solution. Then, the following claims do hold.

 (i) If $u \leq h$ on ∂D, then $u \leq h$ in D.
(ii) If $u \geq h$ on ∂D, then $u \geq h$ in D.

Remark 12.20 Up to replacing u and h by $\Lambda^{-1/2}u$ and $\Lambda^{-1/2}h$ (which are minimizers of $\mathcal{F}_1$ in D), we may assume that $\Lambda = 1$.

We will give two different proofs to Proposition 12.19. The first one is more natural, but is based on the notion of viscosity solution and so it requires the results from Sect. 7.1. The second proof is direct and is based on a purely variational argument in the spirit of Lemma 2.13.

Proof I of Proposition 12.19 By Proposition 7.1, u is a viscosity solution (see Definition 7.6) of

$$\Delta u = 0 \quad \text{in} \quad \Omega_u \cap D, \qquad |\nabla u| = 1 \quad \text{on} \quad \partial \Omega_u \cap D.$$

The conclusion now follows by Lemma 12.21 bellow. $\qquad\qquad\qquad\qquad\qquad$ □

Proof II of Proposition 12.19 We only prove the first claim, the proof of the second one being analogous. For every $t > 0$, consider the half-plane solution

$$h_t(x) = (x \cdot v + c + t)_+.$$

Then, for every $x \in \partial D \cap \Omega_u$, we have that $h(x) \geq u(x) > 0$ and so,

$$u(x) \leq h(x) \leq h_t(x) - t.$$

Thus, we can apply Lemma 12.22 to u and h_t, obtaining that $u \leq h_t$ in D. Since t is arbitrary, we obtain claim (i). $\qquad\qquad\qquad\qquad\qquad\qquad\qquad\qquad$ □

Lemma 12.21 (Comparison of a Viscosity and a Half-Plane Solution) *Let D be a bounded open set in $\mathbb{R}^d$ and let $u : D \to \mathbb{R}$ be a non-negative continuous function and a viscosity solution (see Definition 7.6) to*

$$\Delta u = 0 \quad \text{in} \quad \Omega_u \cap D, \qquad |\nabla u| = 1 \quad \text{on} \quad \partial \Omega_u \cap D.$$

Let $c \in \mathbb{R}$ be a constant, $v \in \mathbb{R}^d$ be a unit vector and $h(x) = (x \cdot v + c)_+$ be a half-plane solution. Then, the following claims do hold.

(i) If $u \geq h$ on ∂D, then $u \geq h$ in D.
(ii) If $u \leq h$ on ∂D, then $u \leq h$ in D.

Proof We first prove (i). Let $M = \|h\|_{L^\infty(D)}$.
For any $t > 0$, we define the real function $f_t : \mathbb{R} \to \mathbb{R}$ as

$$f_t(s) = (1 + t) \max\{s, 0\} + t \left(\max\{s, 0\} \right)^2,$$

for every $s \in \mathbb{R}$. Then, it is immediate to check that the function

$$v_t(x) = f_t \left(x \cdot v + c - M(M + 1)t \right)$$

satisfies the following conditions:

(1) $\Delta v_t > 0$ in the set $\{v_t > 0\}$;
(2) $|\nabla v_t| > 1$ on $\overline{\{v_t > 0\}}$;
(3) $v_t(x) \leq h(x) \leq u(x)$ for every $x \in \partial D$.

Indeed, the first two conditions are immediate, since h is the positive part of an affine function. In order to prove (3), we notice that the inequality is trivial whenever $x \cdot v + c - M(M + 1)t \leq 0$. The case $x \cdot v + c - M(M + 1)t > 0$ is a consequence of the following estimate, which holds for any $S := x \cdot v + c > M(M + 1)t$.

$$f_t(S - M(M + 1)t) = (1 + t)\big(S - M(M + 1)t\big) + t\big(S - M(M + 1)t\big)^2$$
$$\leq (1 + t)\big(S - M(M + 1)t\big) + Mt\big(S - M(M + 1)t\big)$$
$$= S + t\Big(- M(M + 1) + S - M(M + 1)t + MS - M^2(M + 1)t\Big)$$
$$\leq S + t\big(- M(M + 1) + S(M + 1)\big) \leq S.$$

We next claim that $v_t \leq u$ on D. Indeed, suppose that this is not the case and let $T > 0$ be the smallest real number such that $(v_t - T)_+ \leq u$ on $\overline{D}$. Then, there is $x_0 \in \overline{\Omega}_u$ such that $v_t(x_0) - T = u(x_0)$ and $(v_t(x) - T)_+ \leq u(x)$, for every other $x \in \overline{D}$, that is, the test function $(v_t - T)_+$ touches from below u at x_0. Since u is a viscosity solution (see Definition 7.6 and Proposition 7.1) of

$$\Delta u = 0 \quad \text{in} \quad \Omega_u \cap D, \qquad |\nabla u| = 1 \quad \text{on} \quad \partial \Omega_u \cap D,$$

we have that $x_0 \notin \partial \Omega_u \cap B_{3/2}$ and $x_0 \notin \Omega_u$. Then, the only possibility is that $x_0 \in \partial D$, but this is also impossible since $(v_t - T)_+ < v_t \leq u$ on ∂D. This proves that $v_t \leq u$ on D. Now, letting $t \to 0$, we get that

$$u(x) \geq h(x) \quad \text{in} \quad D,$$

which concludes the proof of (i).

The proof of claim (ii) is analogous. We give the proof for the sake of completeness. For any $t > 0$, we define the real function

$$g_t(s) = (1 - \varepsilon t) \max\{s, 0\} - \varepsilon t \left(\max\{s, 0\}\right)^2 \qquad \text{for every} \qquad s \in \mathbb{R},$$

where $\varepsilon > 0$ will be chosen below. We set

$$M_u = \operatorname{diam}(D) + |c| + \|u\|_{L^\infty(D)} \qquad \text{and} \qquad M_h = \|h\|_{L^\infty(D)}.$$

The test function

$$w_t(x) = g_t\big(x \cdot v + c + t\big)$$

satisfies the following conditions:

1. $w_t \geq 0$ for every $0 < t \leq M_u$ and every $s \leq M_h$;
2. $\Delta w_t < 0$ in the open set $\{w_t > 0\}$;
3. $|\nabla w_t| < 1$ on the closed set $\overline{\{w_t > 0\}}$;
4. $w_t(x) \geq h(x) \geq u(x)$ for every $x \in \partial D$ and every $t \leq M_u$.

We start with (1_w). We notice that

$$g_t(s) = (1 - \varepsilon t) \max\{s, 0\} - \varepsilon t \left(\max\{s, 0\}\right)^2 \geq 1 - \varepsilon\left(M_u M_h + M_u M_h^2\right).$$

Thus, in order to have (1_w), we choose

$$\varepsilon \leq \left(M_u M_h + M_u M_h^2\right)^{-1}.$$

Again (2_w) and (3_w) are trivial, while for (4_w) we will need the following estimate, which holds for every $S > 0$ and $t > 0$.

$$\begin{aligned} g_t(S + t) &= (1 - \varepsilon t)(S + t) - \varepsilon t(S + t)^2 \\ &\geq (1 - \varepsilon t)(S + t) - \varepsilon t S(S + t) \\ &= S + t\left(1 - \varepsilon S - \varepsilon t - \varepsilon S^2 - \varepsilon St\right) \\ &\geq S + t\left(1 - \varepsilon\left(S + t + S^2 + ST\right)\right). \end{aligned} \qquad (12.37)$$

In order to have (4_w), we choose

$$\varepsilon \leq \left(M_h + M_u + M_h^2 + M_h M_u\right)^{-1}. \qquad (12.38)$$

We next complete the proof of (4_w). First, notice that the second inequality is always true by hypothesis. Since $w_t \geq 0$, the first inequality is trivial whenever $x \cdot v + c \leq 0$. Thus, we only need to prove that $w_t(x) \geq h(x)$, whenever $x \cdot v + c > 0$. This follows by (12.37) and the second bound on ε (12.38). This concludes the proof of $(1_w) - (4_w)$.

We now consider the set

$$I := \left\{ t \in [0, M_u] \; : \; w_t \geq u \quad \text{on} \quad D \right\}.$$

We notice that I_t is non-empty since $M_u \in I_t$. Let

$$T = \inf I.$$

If $T > 0$, then there is a point $x_0 \in \overline{\Omega}_u$ such that w_T touches u from above in x_0. But this contradicts $(2_w) - (4_w)$. Indeed, (2_w) implies that $x_0 \notin \Omega_u \cap D$, (3_w) gives that $x_0 \notin \partial\Omega_u \cap D$ and (4_w) gives that $x_0 \notin \partial D$. Thus, $T = 0$, which concludes the proof. $\qquad\square$

Lemma 12.22 (Comparison of Minimizers) *Let D be a bounded open set in $\mathbb{R}^d$ and $u, v : \overline{D} \to \mathbb{R}$ be continuous non-negative functions and minimizers of $\mathcal{F}_\Lambda$ in D. Suppose that:*

(a) $u \le v$ on ∂D;
(b) the above inequality is strict on the set $\overline{\Omega}_u \cap \partial D$, that is, $\displaystyle\min_{\overline{\Omega}_u \cap \partial D} (v - u) = m > 0$.

Then, $u \le v$ in D.

Proof Let $\Omega := \{x \in D : u(x) > v(x)\}$. We will prove that $\Omega = \emptyset$. We first claim that Ω is strictly contained in D, that is

$$\partial \Omega \cap \partial D = \emptyset.$$

Suppose that this is not the case. Then, there is a sequence $x_n \in \Omega$ converging to some $x_0 \in \partial D$. Since u and v are continuous, we get that

$$v(x_0) - u(x_0) = 0.$$

On the other hand, for every $n \in \mathbb{N}$, we have

$$u(x_n) > v(x_n) \ge 0,$$

which gives that $x_n \in \Omega_u$. Then, $x_n \in \Omega_u$ and thus, $x_0 \in \partial\Omega_u$. This is a contradiction with the assumption (b).

We will next prove that

$$\Omega_u \cap \partial\{u > v\} = \Omega_v \cap \partial\{u > v\} = \emptyset.$$

We consider the competitors

$$u \vee v = \max\{u, v\} \quad \text{and} \quad u \wedge v = \min\{u, v\}.$$

Since

$$u \vee v = v \quad \text{on} \quad \partial D \quad \text{and} \quad u \wedge v = u \quad \text{on} \quad \partial D,$$

the minimality of u and v implies that

$$\mathcal{F}_\Lambda(v, D) \le \mathcal{F}_\Lambda(u \vee v, D) \quad \text{and} \quad \mathcal{F}_\Lambda(u, D) \le \mathcal{F}_\Lambda(u \wedge v, D). \tag{12.39}$$

On the other hand, we have

$$\mathcal{F}_\Lambda(u \vee v, D) + \mathcal{F}_\Lambda(u \wedge v, D) = \mathcal{F}_\Lambda(u, D) + \mathcal{F}_\Lambda(v, D).$$

Thus, both inequalities in (12.39) are in fact equalities and so $u \wedge v$ is a minimizer of $\mathcal{F}_\Lambda$ in D. Suppose that

$$x_0 \in \Omega_u \cap \partial\Omega.$$

Then, $u(x_0) = v(x_0) > 0$ and by the continuity of u and v, there is a ball $B_r(x_0)$ such that

$$B_r(x_0) \subset \Omega_u \quad \text{and} \quad B_r(x_0) \subset \Omega_v.$$

Thus, both the functions u and $u \wedge v$ are positive and harmonic in $B_r(x_0)$. Thus, the strong maximum principle implies that $u = u \wedge v$ in $B_r(x_0)$. This is contradiction with the assumption that $x_0 \in \partial\{u > v\}$.

We are now in position to prove that $\Omega = \emptyset$. Indeed, suppose that this is not the case. Then, for every $x_0 \in \partial\Omega_u$, we have that $u(x_0) = 0$. Thus, we consider the function

$$\widetilde{u}(x) = \begin{cases} u(x) & \text{if} \quad x \in \overline{D} \setminus \Omega, \\ 0 & \text{if} \quad x \in \Omega. \end{cases}$$

Then, $\widetilde{u} = u$ on ∂D and $\widetilde{u} \in H^1(D)$ (this follows, from instance from the facts that u is Lipschitz continuous on the compact subsets of D and that $\overline{\Omega} \subset D$). Thus, $\widetilde{u}$ is an admissible competitor for u and we have

$$0 \geq \mathcal{F}_\Lambda(u, D) - \mathcal{F}_\Lambda(\widetilde{u}, D) = \int_\Omega |\nabla u|^2 \, dx + \Lambda |\Omega \cap \Omega_u|.$$

In particular,

$$|\Omega| = |\{u > v\}| = |\{u > v\} \cap \{u > 0\}| = |\Omega \cap \Omega_u| = 0,$$

and so, $\Omega = \emptyset$, which concludes the proof.   □

Appendix A
The Epiperimetric Inequality in Dimension Two

In this section we prove the general epiperimetric inequality, which was stated in Theorem 12.3. We show that both the flatness condition from Theorem 12.1 and the non-degeneracy assumption from [49] are unnecessary. We also give an estimate on the H^1 norm of the competitor h, which is useful when one deals with almost-minimizers of the one-phase problem (see for instance [50]).

Theorem A.1 (Epiperimetric Inequality) *There is a constant $\varepsilon > 0$ such that: if $c \in H^1(\partial B_1)$ is a non-negative function on the boundary of the disk $B_1 \subset \mathbb{R}^2$ then, there exists a (non-negative) function $h \in H^1(B_1)$ such that $h = c$ on ∂B_1 and*

$$W(h) - \pi \leq (1 - \varepsilon)\big(W(z) - \pi\big), \tag{A.1}$$

$z \in H^1(B_1)$ being the one-homogeneous extension of c in B_1, that is, $z(x) = |x| c\,(x/|x|)$. Moreover, we can choose the competitor h such that

$$\|h\|_{H^1(B_1)} \leq C\|z\|^2_{H^1(B_1)},$$

where C is a universal numerical constant.

In order to prove Theorem 12.3, we will still use Lemma 12.6, Lemma 12.7 and the results from Sect. 12.1. Moreover, we will need the following results:

Lemma A.2 (Epiperimetric Inequality for Principal Modes: Large Intervals) *Let B_1 be the unit ball in $\mathbb{R}^2$. There is a constant $\varepsilon > 0$ such that: if $c : \partial B_1 \to \mathbb{R}$, $c \in H^1(B_1)$, is a multiple of the first eigenfunction on $\{c > 0\} \subset \partial B_1$ and*

$$\pi \leq \mathcal{H}^1(\{c > 0\} \cap \partial B_1) < 2\pi,$$

© The Author(s) 2023
B. Velichkov, *Regularity of the One-phase Free Boundaries*,
Lecture Notes of the Unione Matematica Italiana 28,
https://doi.org/10.1007/978-3-031-13238-4

then, there exists a (non-negative) function $h \in H^1(B_1)$ such that $h = c$ on ∂B_1 and

$$W(h) - \pi \leq (1 - \varepsilon)\big(W(z) - \pi\big),$$

$z \in H^1(B_1)$ *being the one-homogeneous extension of c in B_1.*

Lemma A.3 (Homogeneity Improvement of the Large Cones) *Let B_1 be the unit ball in $\mathbb{R}^d$, $d \geq 2$. There exist dimensional constants $\eta_0 > 0$ and $\varepsilon > 0$ such that: If $c \in H^1(\partial B_1)$ is non-negative and such that*

$$\frac{1}{d\omega_d}\mathcal{H}^{d-1}(\{c > 0\} \cap \partial B_1) \geq 1 - \eta_0,$$

then we have

$$W(h) - \frac{\omega_d}{2} \leq (1 - \varepsilon)\left(W(z) - \frac{\omega_d}{2}\right),$$

where z is the one-homogeneous extension of c in B_1, while h is given by h_1 or h_2, where

(i) h_1 is the harmonic extension of c in B_1;
(ii) $h_2 : B_1 \to \mathbb{R}$ is given by

$$h_2(r, \theta) = \big(\max\{0, r - \rho\}\big)^\alpha \frac{c(\theta)}{(1 - \rho)^\alpha},$$

where $\alpha > 1$ and $\rho \in (0, 1)$ are dimensional constants.

In both cases, there is a dimensional constant $C_d > 0$ such that

$$\|h\|_{H^1(B_1)} \leq C_d \|c\|_{H^1(\partial B_1)}.$$

A.1   Proof of Theorem 12.3

As in the proof of Theorem 12.1, we decompose the open set $\{c > 0\} \subset \partial B_1$ as a countable union of disjoint arcs, that is,

$$\{c > 0\} = \bigcup_{j \geq 1} \mathcal{I}_j.$$

Fig. A.1 The supports of the one homogeneous extension z (in red) and the competitor h (in blue) in Case 3; the length of the arc $\mathcal{I}_1$ is bigger than $\pi + \delta_0$

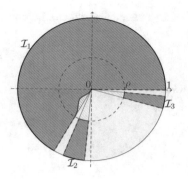

We recall that $|\mathcal{I}_j|$ denotes the length of the arc $\mathcal{I}_j$. Let $\delta_0 > 0$ be a (small) constant that will be chosen later. We consider four cases:

Case 1. There is one arc, say $\mathcal{I}_1$, which is big, that is,

$$\pi - \delta_0 \leq |\mathcal{I}_1| \leq \pi + \delta_0$$

while all the other arcs are small:

$$|\mathcal{I}_j| \leq \pi - \delta_0 \quad \text{for all} \quad j \geq 2.$$

This is precisely Case 1 from the proof of Theorem 12.1 (Sect. 12.4, Fig. 12.2).

Case 2. All the arcs are small, that is,

$$|\mathcal{I}_j| \leq \pi - \delta_0 \quad \text{for all} \quad j \geq 1.$$

This is Case 2 from the proof of Theorem 12.1 (Sect. 12.4, Fig. 12.3).

Case 3 (Fig. A.1). The arc $\mathcal{I}_1$ is very big, that is,

$$\pi + \delta_0 \leq |\mathcal{I}_1| \leq 2\pi - \delta_0.$$

As a consequence, the other arcs are small:

$$|\mathcal{I}_j| \leq \pi - \delta_0 \quad \text{for all} \quad j \geq 2.$$

Case 4. The support of c is very big, that is,

$$2\pi - 2\delta_0 \leq \mathcal{H}^1(\{c > 0\}) \leq 2\pi.$$

In this case the competitor is given precisely by Lemma A.3.

Thus, it is sufficient to consider Case 3. We argue precisely as in the proof of Case 1. Let $\{\phi_j\}_{j \geq 1}$ be a complete orthonormal system of eigenfunctions on $\mathcal{I}_1$, and let c_j be the Fourier coefficient

$$c_j := \int_{\partial B_1} c(\theta)\phi_j(\theta)\, d\theta \qquad \text{for every} \quad j \geq 2.$$

We decompose the trace c as

$$c(\theta) = c_1\phi_1(\theta) + \psi_1(\theta) + \psi_2(\theta),$$

where

$$\psi_1(\theta) = \sum_{j=2}^{\infty} c_j\phi_j(\theta),$$

and ψ_2 is the restriction of c on the set $\bigcup_{j\geq 2} \mathcal{I}_j$. Since ψ_2 is supported on $\bigcup_{j\geq 2} \mathcal{I}_j$ and $|\mathcal{I}_j| \leq \pi - \delta_0$, for $j \geq 2$, we have that

$$\int_{\mathbb{S}^1} |\nabla_\theta \psi_2|^2\, d\theta \geq \frac{\pi^2}{(\pi - \delta_0)^2} \int_{\mathbb{S}^1} \psi_2^2\, d\theta.$$

For what concerns ψ_1, since its Fourier expansion contains only higher modes on $\mathcal{I}_1$ and since $|\mathcal{I}_1| \leq 2\pi - \delta$, we obtain

$$\int_{\mathcal{I}_1} |\nabla_\theta \psi_1|^2\, d\theta \geq \frac{4\pi^2}{(2\pi - \delta_0)^2} \int_{\mathcal{I}_1} \psi_1^2\, d\theta.$$

Thus, there is κ depending only on δ_0 such that

$$\int_{\mathbb{S}^1} |\nabla_\theta \psi_i|^2\, d\theta \geq (1 + \kappa) \int_{\mathbb{S}^1} \psi_i^2\, d\theta \qquad \text{for} \quad i = 1, 2.$$

Let $\rho > 0$ and $\varepsilon_\psi > 0$ be the constants from Lemma 12.6 corresponding to the constant κ from (12.19); let h_{ψ_1} and h_{ψ_2} be the competitors from Lemma 12.6 associated to the traces ψ_1 and ψ_2, respectively. Thus, setting $z_{\psi_i}(r, \theta) := z\psi_i(\theta)$, we have

$$W_0(h_{\psi_1}) \leq (1 - \varepsilon_\psi) W_0(z_{\psi_1}) \qquad \text{and} \qquad W(h_{\psi_2}) \leq (1 - \varepsilon_\psi) W(z_{\psi_2}). \qquad (A.2)$$

Let $\tilde{h}$ be the competitor from Lemma A.3 with trace $c_1\phi_1$ and let $\tilde{z}(r, \theta) := rc_1\phi_1(\theta)$. We set

$$\tilde{h}_\rho(r, \theta) = \begin{cases} \tilde{z}(r, \theta) & \text{if} \quad r \in [\rho, 1], \\ \rho\, \tilde{h}(r/\rho, \theta) & \text{if} \quad r \in [0, \rho]. \end{cases}$$

Thus, Lemma 12.7 and Lemma 2.3 imply that

$$W(\tilde{h}_\rho) - \pi \leq (1 - \rho^d \tilde{\varepsilon})(W(\tilde{z}) - \pi), \tag{A.3}$$

$\tilde{\varepsilon}$ being the constant from Lemma A.3. Finally, we define the competitor $h : B_1 \to \mathbb{R}$ as:

- $h = z$ if $W(z) \leq \pi$, where $z = \tilde{z} + z_{\psi_1} + z_{\psi_2}$ is the 1-homogeneous extension of c;
- $h = \tilde{z} + h_{\psi_1} + h_{\psi_2}$ if $W(z) > \pi$, but $W(\tilde{z}) \leq \pi$;
- $h = \tilde{h} + h_{\psi_1} + h_{\psi_2}$ if $W(z) > \pi$ and $W(\tilde{z}) > \pi$.

Notice that the only non-trivial case is the third one: $W(z) > \pi$ and $W(\tilde{z}) > \pi$. By the decomposition lemma (Lemma 12.9), we have

$$W(z) = W(\tilde{z}) + W_0(z_{\psi_1}) + W(z_{\psi_1}),$$

and

$$W(h) = W(\tilde{h}_\rho) + W_0(h_{\psi_1}) + W(h_{\psi_1}).$$

Setting

$$\varepsilon = \min\{\rho^d \tilde{\varepsilon}, \varepsilon_\psi\},$$

we obtain the epiperimetric inequality (A.1) as a consequence of (A.2) and (A.3). This concludes the proof in Case 3. $\qquad\square$

A.2   Proof of Lemma A.2

We will use the notations from Sect. 12.3 In this case, we have that $\delta \in (0, \pi)$. The infinitesimal argument used in the proof of Lemma 12.7 cannot be applied here. Thus, we directly compute $F(\delta)$ (defined in (12.15)) by using the identity from Sect. 12.3.3.

$$
\begin{aligned}
F(\delta) &= \int_0^1 \left(\frac{(t(r)/\pi)^2}{1 + t(r)/\pi} - (1 - \varepsilon) \frac{(\delta/\pi)^2}{1 + \delta/\pi} \right) r\,dr \\
&= \int_0^1 \frac{(\pi + \delta)t(r)^2 - (\pi + t(r))(1 - \varepsilon)\delta^2}{\pi^3 \, (1 + t(r)/\pi) \, (1 + \delta/\pi)} r\,dr \\
&\leq \frac{\delta^2}{\pi^3} \int_0^1 \left((\pi + \delta)\,(t(r)/\delta)^2 - (\pi + t(r))(1 - \varepsilon) \right) r\,dr.
\end{aligned}
$$

Now, since $t(r)/\delta = 1 - 3(1-r)\varepsilon$, we get that

$$(\pi + \delta)\left(t(r)/\delta\right)^2 - (\pi + t(r))(1 - \varepsilon)$$

$$= (\pi + \delta)\left(1 - 3(1-r)\varepsilon\right)^2 - \left(\pi + \delta - 3(1-r)\varepsilon\delta\right)(1 - \varepsilon)$$

$$= -\varepsilon(\pi + \delta)(5 - 6r) + 9\varepsilon^2(\pi + \delta)(1 - r)^2 + 3(1-r)\varepsilon\delta(1 - \varepsilon)$$

$$\leq -\varepsilon(\pi + \delta)(5 - 6r) + 18\pi\varepsilon^2 + 3(1-r)\varepsilon\delta.$$

Thus, multiplying by r and integrating over $[0, 1]$, we get

$$F(\delta) \leq \frac{\delta^2}{\pi^3}\left(-\frac{1}{2}\varepsilon(\pi + \delta) + 9\pi\varepsilon^2 + \frac{1}{2}\varepsilon\delta\right) = -\frac{\varepsilon\delta^2}{2\pi^2}\left(1 - 18\varepsilon\right).$$

Thus, using (12.14), we get

$$\left(W(h_\delta) - \frac{\pi}{2}\right) - (1 - \varepsilon)\left(W(z_\delta) - \frac{\pi}{2}\right) \leq -\frac{\varepsilon\delta^2}{2\pi^2}\left(1 - 18\varepsilon - 2\pi^2 C\varepsilon\right).$$

Choosing $\varepsilon > 0$ small enough, we get (12.13). This concludes the proof of Lemma A.2.   □

A.3   Epiperimetric Inequality for Large Cones: Proof of Lemma A.3

We write the trace c in Fourier series on the sphere ∂B_1 as

$$c(\theta) = c_0\phi_0 + c_1\phi_1(\theta) + \sum_{j=2}^{\infty} c_j\phi_j(\theta),$$

where:

- ϕ_0 is the constant $(d\omega_d)^{-1/2}$;
- $\phi_1 : \partial B_1 \to \mathbb{R}$ is an eigenfunction of the Laplacian on the sphere, the respective eigenvalue being $(d - 1)$ and $\int_{\partial B_1} \phi_1^2 \, d\theta = 1$;
- ϕ_j, for $j \geq 2$ are eigenfunctions orthonormal in $L^2(\partial B_1)$ with eigenvalues $\lambda_j \geq 2d$ on the sphere ∂B_1.

We now set

$$\psi(\theta) := \sum_{j=2}^{\infty} c_j\phi_j(\theta).$$

Since the Fourier expansion of ψ contains only eigenfunctions associated to eigenvalues $\geq 2d$, we get that

$$\int_{\mathbb{S}^{d-1}} |\nabla_\theta \psi|^2 \, d\theta \geq 2d \int_{\mathbb{S}^{d-1}} \psi^2 \, d\theta.$$

Let $\kappa = d + 1$ and ρ_κ, ε_κ and α_κ be the constants from Lemma 12.6; let $z_\kappa(r, \theta) = r\psi(\theta)$ and h_κ be the competitor from Lemma 12.6. We choose

$$\rho = \rho_\kappa \qquad \text{and} \qquad \alpha = \alpha_\kappa.$$

We consider the functions

$$z(r, \theta) = c_0 r \phi_0 + c_1 r \phi_1(\theta) + z_\kappa(r, \theta)$$

$$\tilde{h}_1(r, \theta) = c_0 \phi_0 + r c_1 \phi_1(\theta) + h_\kappa(r, \theta).$$

$$h_2(r, \theta) = \frac{(r - \rho)_+^\alpha}{(1 - \rho)^\alpha} c_0 \phi_0 + \frac{(r - \rho)_+^\alpha}{(1 - \rho)^\alpha} c_1 \phi_1(\theta) + h_\kappa(r, \theta).$$

Step 1. We first calculate the terms

$$W_0(\phi_0), \quad W_0(r\phi_0), \quad W_0(r\phi_1(\theta)), \quad W_0((r - \rho)_+^\alpha \phi_0) \quad \text{and} \quad W_0((r - \rho)_+^\alpha \phi_1(\theta)).$$

Since ϕ_0 is a constant, we have that

$$W_0(\phi_0) = -\int_{\partial B_1} \phi_0^2 \, d\theta = -1.$$

Since $r\phi_1(\theta)$ is one homogeneous, we get

$$W_0(r\phi_0) = \frac{1}{d} \mathcal{F}_0(\phi_0) = -\frac{d-1}{d} \int_{\partial B_1} \phi_0^2 \, d\theta = -\frac{d-1}{d}.$$

Analogously, we obtain

$$W_0(r\phi_1) = \frac{1}{d} \mathcal{F}_0(\phi_1) = \frac{1}{d} \left(\int_{\partial B_1} |\nabla_\theta \phi_1|^2 \, d\theta - (d-1) \int_{\partial B_1} \phi_1^2 \, d\theta \right) = 0,$$

since ϕ_1 is a $(d-1)$-eigenfunction on the sphere. For the last two terms, we use the formula

$$
W_0\big((r-\rho)^\alpha_+ \phi_i(\theta)\big) = \int_\rho^1 r^{d-1}\, dr \int_{\partial B_1} \left(\alpha^2 (r-\rho)^{2\alpha-2} \phi_i^2 + \frac{(r-\rho)^{2\alpha}}{r^2} |\nabla_\theta \phi_i|^2 \right) d\theta
$$

$$
- (1-\rho)^{2\alpha} \int_{\partial B_1} \phi_i^2\, d\theta
$$

$$
\leq \frac{1}{2\alpha+d-2} \left(\alpha^2 + \int_{\partial B_1} |\nabla_\theta \phi_i|^2\, d\theta \right) - (1-\rho)^{2\alpha}.
$$

Since $1 < \alpha \leq 2$, $\displaystyle\int_{\partial B_1} |\nabla_\theta \phi_0|^2\, d\theta = 0$ and $\displaystyle\int_{\partial B_1} |\nabla_\theta \phi_1|^2\, d\theta = d-1$, we get

$$
W_0\big((r-\rho)_+\phi_0^\alpha\big) \leq 1 - (1-\rho)^{2\alpha} \qquad \text{and} \qquad W_0\big((r-\rho)^\alpha_+\phi_1(\theta)\big) \leq 2 - (1-\rho)^{2\alpha}.
$$

Step 2. Consider the competitor h_1. We set

$$
\eta := \frac{1}{d\omega_d} \mathcal{H}^{d-1}(\{c=0\} \cap \partial B_1),
$$

and we calculate

$$
W(h_1) - W(z) \leq W_0(h_1) - W_0(z) + \frac{1}{d}\eta \leq W_0(\tilde{h}_1) - W_0(z) + \omega_d \eta
$$

$$
= \left(c_0^2 W_0(\phi_0) + c_1^2 W_0\big(r\phi_1(\theta)\big) + W_0(h_k) \right)
$$

$$
- \left(c_0^2 W_0(r\phi_0) + c_1^2 W_0(r\phi_1) + W_0(z_\kappa) \right) + \omega_d \eta
$$

$$
= c_0^2 \Big(W_0(\phi_0) - W_0(r\phi_0) \Big) + \Big(W_0(h_k) - W_0(z_\kappa) \Big)
$$

$$
+ \frac{\eta}{d} \leq -\frac{c_0^2}{d} - \varepsilon_\kappa W_0(z_\kappa) + \omega_d \eta.
$$

Step 3. For the competitor h_2 we calculate

$$
W(h_2) - W(z) = W_0(h_2) - W_0(z) + |\{h_2 > 0\} \cap B_1| - |\{z > 0\} \cap B_1|
$$

$$
= c_0^2 \left(\frac{1}{(1-\rho)^{2\alpha}} W_0\big((r-\rho)^\alpha_+\phi_0\big) - W_0(r\phi_0) \right)
$$

$$
+ c_1^2 \left(\frac{1}{(1-\rho)^{2\alpha}} W_0\big((r-\rho)^\alpha_+\phi_1\big) - W_0(r\phi_1) \right)
$$

$$+ \left(W_0(h_k) - W_0(z_\kappa) \right) - |\{z > 0\} \cap B_\rho|$$

$$\leq \frac{2}{(1 - \rho)^{2\alpha}} (c_0^2 + c_1^2) - \varepsilon_k W_0(z_\kappa) - \omega_d \rho^d (1 - \eta) .$$

Now, since

$$|c_1| = \left| \int_{B_1} c(\theta) \phi_1 \, d\theta \right| \leq \|\phi_1\|_{L^\infty(\partial B_1)} \int_{B_1} c(\theta) \, d\theta \leq \|\phi_1\|_{L^\infty(\partial B_1)} d\omega_d c_0,$$

there is a dimensional constant $C_d > 0$ such that

$$W(h_2^+) - W(z) \leq C_d \, c_0^2 - \varepsilon_k W_0(z_\kappa) - \omega_d \rho^d (1 - \eta) .$$

Step 4. Conclusion of the proof. We calculate the energy

$$W(z) - \frac{\omega_d}{2} = c_0^2 W_0(r\phi_0) + c_1^2 W_0(r\phi_1) + W_0(z_\kappa) + \frac{\omega_d}{2} - \omega_d \eta$$

$$= -\frac{d - 1}{d} c_0^2 + W_0(z_\kappa) + \omega_d (1/2 - \eta) .$$

Since $W_0(z_\kappa) > 0$, for every $\varepsilon \in (0, \varepsilon_\kappa)$, we have that the inequality

$$-\frac{c_0^2}{d} + \omega_d \eta \leq \varepsilon \left(\frac{d - 1}{d} c_0^2 - \omega_d (1/2 - \eta) \right), \tag{A.4}$$

implies that

$$W(h_1) - W(z) \leq -\varepsilon \left(W(z) - \frac{\omega_d}{2} \right).$$

Analogously,

$$C_d \, c_0^2 - \omega_d \rho^d (1 - \eta) \leq \varepsilon \left(\frac{d - 1}{d} c_0^2 - \omega_d (1/2 - \eta) \right), \tag{A.5}$$

implies that

$$W(h_2) - W(z) \leq -\varepsilon \left(W(z) - \frac{\omega_d}{2} \right).$$

Now, (A.4) is implied by

$$d\omega_d (\eta + \varepsilon) \leq c_0^2. \tag{A.6}$$

while if we assume that $\eta_0 \leq 1/4$, we get that (A.5) is implied by

$$C_d c_0^2 \leq \frac{\omega_d}{2}(\rho^d - \varepsilon). \tag{A.7}$$

Now, if both (A.6) and (A.7) were false, then we would have

$$d\omega_d(\eta + \varepsilon)C_d \geq C_d c_0^2 \geq \frac{\omega_d}{2}(\rho^d - \varepsilon),$$

and finally

$$\eta \geq \frac{1}{2dC_d}\left(\rho^d - \varepsilon(1 + 2dC_d)\right), \tag{A.8}$$

Finally, we notice that the choice

$$\varepsilon = \frac{\rho^d}{4dC_d + 2} \qquad \text{and} \qquad \eta_0 = \frac{\rho^d}{2dC_d},$$

makes (A.8) impossible and concludes the proof of Lemma A.3.   □

Appendix B
Notations and Definitions

Euclidean Space, Topology and Distance

$$\mathbb{R}^d \quad | \quad x \cdot y \quad | \quad \text{dist}(x, K) \quad | \quad \text{dist}_{\mathcal{H}}(K_1, K_2) \quad | \quad \text{diam } K \quad | \quad B_r(x) \quad | \quad \bar{\Omega} \quad | \quad \partial \Omega$$

- d is the dimension of the space.
- C_d denotes a positive constant that depends only on the dimension;
  C_d may change from line to line and even within the same line.
- $x = (x_1, \ldots, x_d)$ denotes a generic point in $\mathbb{R}^d$; we will also write

$$x = (x', x_d), \text{ where } x' = (x_1, \ldots, x_{d-1}) \text{ is a point in } \mathbb{R}^{d-1}.$$

- We denote by $x \cdot y := \sum_{i=1}^{d} x_i y_i$ the scalar product of two vectors $x = (x_1, \ldots, x_d)$ and $y = (y_1, \ldots, y_d)$ in $\mathbb{R}^d$; $|x| = \sqrt{x \cdot x}$ is the euclidean norm of x in $\mathbb{R}^d$.
- The standard orthonormal basis of $\mathbb{R}^d$ will be denoted by $\{e_1, \ldots, e_d\}$;
  e_d is the unit vector with coordinates $(0, \ldots, 0, 1)$.
- By $\text{dist}(x, K)$ we denote the euclidean distance from a point $x \in \mathbb{R}^d$ to a set $K \subset \mathbb{R}^d$

$$\text{dist}(x, K) = \min_{y \in K} |x - y|.$$

- Given two sets K_1 and K_2 in $\mathbb{R}^d$, we denote by $\text{dist}_{\mathcal{H}}(K_1, K_2)$ the Hausdorff distance between K_1 and K_2, that is,

$$\text{dist}_{\mathcal{H}}(K_1, K_2) := \max \left\{ \max_{x \in K_1} \text{dist}(x, K_2), \max_{y \in K_2} \text{dist}(y, K_1) \right\}.$$

© The Author(s) 2023
B. Velichkov, *Regularity of the One-phase Free Boundaries*,
Lecture Notes of the Unione Matematica Italiana 28,
https://doi.org/10.1007/978-3-031-13238-4

- diam K stands for the diameter of a set $K \subset \mathbb{R}^d$

$$\text{diam } K := \sup \{ |x - y| : x, y \in K \}.$$

- $B_r(x)$ is the ball of radius r and center x; B_r is the ball of radius r and center 0.
- For any set $\Omega \subset \mathbb{R}^d$, we denote by $\bar{\Omega}$ its closure and by $\partial \Omega$ its boundary;

Measure and Integration

$$| \Omega | \quad | \quad \omega_d \quad | \quad \Omega^{(\alpha)} \quad | \quad \mathcal{H}^s_\delta \quad | \quad \mathcal{H}^s \quad | \quad \mathcal{H}^{d-1} \quad | \quad \dim_{\mathcal{H}} \quad | \quad f\!\!\!-$$

- $|\Omega|$ is the Lebesgue measure of a (Lebesgue measurable) set $\Omega \subset \mathbb{R}^d$.
- By ω_d we denote the Lebesgue measure of the unit ball in $\mathbb{R}^d$.
- For any $\Omega \subset \mathbb{R}^d$ and $\alpha \in [0, 1]$, we define the set $\Omega^{(\alpha)}$ as the set of points at which Ω has Lebesgue density equal to α, that is,

$$\Omega^{(\alpha)} = \left\{ x_0 \in \mathbb{R}^d : \lim_{r \to 0} \frac{|\Omega \cap B_r(x_0)|}{|B_r|} = \alpha \right\}.$$

We recall that

$$|\Omega \setminus \Omega^{(1)}| = |\Omega^{(1)} \setminus \Omega| = 0 \quad \text{and} \quad |\Omega \cap \Omega^{(0)}| = 0.$$

- For every $s > 0$, $\delta \in (0, +\infty]$ and every set $E \subset \mathbb{R}^d$, we define

$$\mathcal{H}^s_\delta(E) := \frac{\omega_s}{2^s} \inf \left\{ \sum_{j=1}^\infty \left(\text{diam } U_j \right)^s : \text{for every family of sets } \{U_j\}_{j=1}^\infty \right.$$
$$\left. \text{such that } E \subset \bigcup_{j=1}^\infty U_j \text{ and diam } U_j \leq \delta \right\},$$

where, if $s \in \mathbb{N}$, then ω_s is the measure of the unit ball in $\mathbb{R}^s$, and we recall that ω_s can be defined for any $s \in (0, +\infty)$ as

$$\omega_s := \frac{\pi^{s/2}}{\Gamma(s/2 + 1)} \quad \text{where} \quad \Gamma(s) := \int_0^{+\infty} x^{s-1} e^x \, dx.$$

- For any $s \geq 0$, $\mathcal{H}^s(E)$ denotes the s-dimensional Hausdorff measure of a set $E \subset \mathbb{R}^d$.

$$\mathcal{H}^s(E) = \lim_{\delta \to 0_+} \mathcal{H}^s_\delta(E) = \sup_{\delta > 0} \mathcal{H}^s_\delta(E).$$

For instance, we have

$$\mathcal{H}^d(B_r) = |B_r| = \omega_d r^d \qquad \text{and} \qquad \mathcal{H}^{d-1}(\partial B_r) = d\omega_d r^{d-1}.$$

- The Hausdorff dimension of a set $E \subset \mathbb{R}^d$ is defined as

$$\dim_{\mathcal{H}} E = \inf\{s > 0 \ : \ \mathcal{H}^s(E) = 0\}.$$

- By $\fint_{\Omega} \phi \, d\mu$ we will indicate the mean value of the function ϕ on the set Ω with respect to the measure μ, that is, $\fint_{\Omega} \phi \, d\mu := \dfrac{1}{\mu(\Omega)} \int_{\Omega} \phi \, d\mu.$

Perimeter and Reduced Boundary

$$\boxed{\quad \partial^* \Omega \quad | \quad Per(\Omega) \quad | \quad Per(\Omega; D) \quad}$$

- Let $\Omega \subset \mathbb{R}^d$ be a Lebesgue measurable set in $\mathbb{R}^d$. We say that Ω is a set of finite perimeter (in the sense of De Giorgi) if

$$Per(\Omega) := \sup\left\{ \int_{\Omega} \operatorname{div} \xi \, dx \ : \ \xi \in C_c^1(\mathbb{R}^d; \mathbb{R}^d), \ \|\xi\|_{L^\infty(\mathbb{R}^d)} \leq 1 \right\} < +\infty.$$

Analogously, we define the relative perimeter of Ω in the open set $D \subset \mathbb{R}^d$ as

$$Per(\Omega; D) := \sup\left\{ \int_{\Omega} \operatorname{div} \xi \, dx \ : \ \xi \in C_c^1(D; \mathbb{R}^d), \ \|\xi\|_{L^\infty(D)} \leq 1 \right\}.$$

- Equivalently, $\Omega \subset \mathbb{R}^d$ is a set of finite perimeter if there is an $\mathbb{R}^d$-valued Radon measure μ_Ω such that

$$\int_{\Omega} \nabla\phi(x) \, dx = \int_{\mathbb{R}^d} \phi \, d\mu_\Omega \qquad \text{for every} \qquad \phi \in C_c^1(\mathbb{R}^d).$$

With this notations, we have

$$Per(\Omega) = |\mu_\Omega|(\mathbb{R}^d) \qquad \text{and} \qquad Per(\Omega; D) = |\mu_\Omega|(D),$$

where by $|\mu|$ we denote the total variation of a measure μ.

- The reduced boundary $\partial^*\Omega$ is defined as

$$\partial^*\Omega := \left\{ x \in \mathbb{R}^d \ : \ \text{the limit } \nu_\Omega(x) := \lim_{r \to 0} \frac{\mu_\Omega(B_r(x))}{|\mu_\Omega|(B_r(x))} \text{ exists and } |\nu_\Omega(x)| = 1 \right\};$$

ν_Ω is called a measure theoretic outer normal at x. The following are well-known facts about the reduced boundary of a set of finite perimeter (we refer to the recent book [43], which is an excellent introduction to this topic):

1. $\mu_\Omega = \nu_\Omega \, \mathcal{H}^{d-1} \lfloor \partial^* \Omega$;
2. $\partial^* \Omega \subset \Omega^{(1/2)}$;
3. setting

$$\Omega_{x,r} := \frac{1}{r}(\Omega - x) = \left\{ \frac{y - x}{r} \; : \; y \in \Omega \right\},$$

we have that the characteristic functions $\mathbb{1}_{\Omega_{x,r}}$ converge (as $r \to 0$) in $L^1_{loc}(\mathbb{R}^d)$ to the characteristic function of the half-space $\{ y \in \mathbb{R}^d \; : \; y \cdot \nu_\Omega(x) < 0 \}$;

(4) $\mathcal{H}^{d-1}\left(\mathbb{R}^d \setminus \left(\Omega^{(1)} \cup \Omega^{(0)} \cup \partial^* \Omega \right) \right) = 0.$

Unit Sphere and Polar Coordinates

$$\boxed{\quad \mathbb{S}^{d-1} \quad | \quad \theta \quad | \quad \nabla_\theta \quad | \quad \Delta_{\mathbb{S}} \quad | \quad d\theta \quad}$$

- $\mathbb{S}^{d-1}$ is the unit $(d-1)$-dimensional sphere; we will indicate by θ the points on $\mathbb{S}^{d-1}$ and we will often identify $\mathbb{S}^{d-1}$ with ∂B_1, where B_1 is the unit ball in $\mathbb{R}^d$; we will sometimes use $d\theta$ to indicate the surface measure on the sphere, thus

$$\int_{\mathbb{S}^{d-1}} \phi(\theta) \, d\theta \,, \quad \int_{\partial B_1} \phi(\theta) \, d\theta \quad \text{and} \quad \int_{\partial B_1} \phi \, d\mathcal{H}^{d-1}$$

all denote the integral of the function $\phi : \partial B_1 \to \mathbb{R}$ on the unit sphere ∂B_1 in $\mathbb{R}^d$.
- For a function $\phi : \partial B_1 \to \mathbb{R}$, we denote by $\nabla_\theta \phi$ its gradient on the sphere ∂B_1.
- We denote by $H^1(\partial B_1)$ the Sobolev space of functions (on the sphere) which are square integrable and have a square integrable gradient. Precisely, $H^1(\partial B_1)$ is the closure of $C^\infty(\partial B_1)$ with respect to the norm

$$\|\phi\|_{H^1(\partial B_1)} := \left(\|\phi\|^2_{H^1(\partial B_1)} + \|\nabla_\theta \phi\|^2_{H^1(\partial B_1)} \right)^{1/2}.$$

- By $\Delta_{\mathbb{S}} \phi$ we denote the (distributional) spherical Laplacian of a Sobolev function $\phi \in H^1(\partial B_1)$; we have the following integration by parts formula

$$\int_{\partial B_1} \psi \Delta_{\mathbb{S}} \phi \, d\theta = - \int_{\partial B_1} \nabla_\theta \psi \cdot \nabla_\theta \phi \, d\theta \quad \text{for every} \quad \psi \in H^1(\partial B_1).$$

- If $u : B_R \to \mathbb{R}$ is a function expressed in polar coordinates as $u = u(r, \theta)$, then

$$|\nabla u|^2 = (\partial_r u)^2 + r^{-2}|\nabla_\theta u|^2,$$

and

$$\Delta u = r^{1-d}\partial_r\left(r^{d-1}\partial_r u\right) + r^{-2}\Delta_{\mathbb{S}} u = \partial_{rr}u + \frac{d-1}{r}\partial_r u + \frac{1}{r^2}\Delta_{\mathbb{S}} u.$$

Moreover, we recall that

$$\int_{B_R} u \, dx = \int_0^R \int_{\mathbb{S}^{d-1}} u(r, \theta) \, d\theta \, r^{d-1} \, dr.$$

Functions and Sets

$$u \wedge v \quad | \quad u \vee v \quad | \quad u_+ \quad | \quad u_- \quad | \quad \{u > 0\} \quad | \quad \Omega_u \quad | \quad \Omega_u^+ \quad | \quad \Omega_u^- \quad | \quad \mathbb{1}_\Omega$$

- Given two real-valued functions u and v defined on the same domain, we denote by $u \wedge v$ and $u \vee v$ the functions

$$(u \wedge v)(x) = \min\{u(x), v(x)\} \qquad \text{and} \qquad (u \vee v)(x) = \max\{u(x), v(x)\}.$$

- $u_+ = u \vee 0$ and $u_- = (-u) \vee 0$. Thus, we have $u = u_+ - u_-$ and $|u| = u_+ + u_-$.

  We do not distinguish between u_+ and u^+, nor between u_- and u^-.

- By $\{u > 0\}$ we mean the set $\{x \in \mathbb{R}^d \ : \ u(x) > 0\}$; the sets $\{u = 0\}$, $\{u \neq 0\}$ and $\{u < 0\}$ are defined analogously. For any u, we set

$$\Omega_u = \{u \neq 0\}, \quad \Omega_u^+ = \{u > 0\} \quad \text{and} \quad \Omega_u^- = \{u < 0\}.$$

- By $\mathbb{1}_\Omega$ we denote the characteristic functions of the set Ω, that is,

$$\mathbb{1}_\Omega(x) = \begin{cases} 1 & \text{if} \quad x \in \Omega, \\ 0 & \text{if} \quad x \notin \Omega. \end{cases}$$

The One-Phase Functional and Related Quantities

$$u_{r,x_0} \quad | \quad \mathcal{F}_\Lambda(u,D) \quad | \quad Reg(\partial\Omega_u) \quad | \quad Sing(\partial\Omega_u) \quad | \quad W(u) \quad | \quad W_0(u) \quad | \quad \delta\mathcal{F}_\Lambda(u,D)[\xi]$$

- For any $r > 0$ and $x_0 \in \mathbb{R}^d$, we denote by u_{r,x_0} and u_r the functions

$$u_{r,x_0}(x) = \frac{1}{r}u(x_0 + rx) \qquad \text{and} \qquad u_r(x) = \frac{1}{r}u(rx).$$

- For any constant $\Lambda \geq 0$, open set $D \subset \mathbb{R}^d$ and function $u \in H^1(D)$, the one-phase functional $\mathcal{F}_\Lambda(u,D)$ is defined as

$$\mathcal{F}_\Lambda(u,D) = \int_D |\nabla u|^2\, dx + \Lambda|\{u > 0\} \cap D|.$$

- The so-called *regular part* $Reg(\partial\Omega_u)$ of the free boundary $\partial\Omega_u$ (see Sect. 6.4) is defined as the set of points $x_0 \in \partial\Omega_u$, for which there exist:

  - an infinitesimal sequence $r_n \to 0$;
  - a unit vector $\nu \in \mathbb{R}^d$;

  such that the blow-up sequence

$$u_n : B_1 \to \mathbb{R}, \qquad u_n(x) = \frac{1}{r_n}u(x_0 + r_n x), \tag{B.1}$$

  converges uniformly in B_1 to a blow-up limit

$$h_\nu : B_1 \to \mathbb{R}, \qquad h_\nu(x) = \sqrt{\Lambda}\,(x \cdot \nu)_+. \tag{B.2}$$

- The singular part $Sing(\partial\Omega_u)$ of the free boundary $\partial\Omega_u$ is defined simply as the complementary of $Reg(\partial\Omega_u)$

$$Sing(\partial\Omega_u) = \partial\Omega_u \setminus Reg(\partial\Omega_u).$$

  For some fine results on the structure of the singular set we refer to Sect. 10.

- By W_Λ we denote the Weiss' boundary adjusted energy (in the ball B_1), that is, for every $u \in H^1(B_1)$, we set

$$W_0(u) = \int_{B_1} |\nabla u|^2\, dx - \int_{\partial B_1} u^2\, d\mathcal{H}^{d-1} \quad \text{and} \quad W_\Lambda(u) = W_0(u) + \Lambda|\{u > 0\} \cap B_1|.$$

  For the related Weiss monotonicity formula see Lemma 9.2. Only in Sect. 12 and in the Appendix A, we use the shorter notation $W := W_1$.

- Let D be an open subset of $\mathbb{R}^d$ and $u \in H^1(D)$ be non-negative. By $\delta\mathcal{F}_\Lambda(u, D)[\xi]$ we denote the first variation of the functional $\mathcal{F}_\Lambda(\cdot, D)$ (calculated at u) in the direction of the compactly supported smooth vector field $\xi \in C_c^\infty(D; \mathbb{R}^d)$. Precisely,

$$\delta\mathcal{F}_\Lambda(u, D)[\xi] = \frac{\partial}{\partial t}\Big|_{t=0} \mathcal{F}_\Lambda(u \circ \Psi_t^{-1}, D),$$

where $\Psi_t(x) = x + t\xi(x)$.

Remark An explicit formula (9.6) for the first variation is given in Lemma 9.5.

Definition We say that u is stationary for $\mathcal{F}_\Lambda$ in D (see Sect. 9.5) if

$$\delta\mathcal{F}_\Lambda(u, D)[\xi] = 0 \quad \text{for every} \quad \xi \in C_c^\infty(D; \mathbb{R}^d).$$

References

1. N. Aguilera, H.W. Alt, L.A. Caffarelli. An optimization problem with volume constraint. SIAM J. Control Optim. **24**(2), 191–198 (1986)
2. F.J. Almgren Jr., Dirichlet's problem for multiple valued functions and the regularity of mass minimizing integral currents. *Minimal Submanifolds and Geodesics* (Kaigai, Tokyo, 1979), pp. 1–6
3. H.W. Alt, L.A. Caffarelli, Existence and regularity for a minimum problem with free boundary. J. Reine Angew. Math. **325**, 105–144 (1981)
4. H.W. Alt, L.A. Caffarelli, A. Friedman, Variational problems with two phases and their free boundaries. Trans. Amer. Math. Soc. **282**(2), 431–461 (1984)
5. T. Briançon, Regularity of optimal shapes for the Dirichlet's energy with volume constraint. ESAIM, Control Optim. Calc. Var. **10**, 99–122 (2004)
6. T. Briançon, J. Lamboley, Regularity of the optimal shape for the first eigenvalue of the Laplacian with volume and inclusion constraints. Ann. Inst. H. Poincaré Anal. Non Linéaire **26**(4), 1149–1163 (2009)
7. T. Briançon, M. Hayouni, M. Pierre, Lipschitz continuity of state functions in some optimal shaping. Calc. Var. PDE **23**(1), 13–32 (2005)
8. D. Bucur, Minimization of the k-th eigenvalue of the Dirichlet Laplacian. Arch. Rat. Mech. Anal. **206**(3), 1073–1083 (2012)
9. D. Bucur, D. Mazzoleni, A. Pratelli, B. Velichkov, Lipschitz regularity of the eigenfunctions on optimal domains. Arch. Ration. Mech. Anal. **216**(1), 117–151 (2015)
10. G. Buttazzo, B. Velichkov, A shape optimal control problem with changing sign data. SIAM J. Math. Anal. **50**(3), 2608–2627 (2018)
11. L. Caffarelli, The regularity of free boundaries in higher dimensions. Acta Math. **139** (1977)
12. L. Caffarelli, A Harnack inequality approach to the regularity of free boundaries. I. Lipschitz free boundaries are $C^{1,\alpha}$. Rev. Mat. Iberoamericana **3**(2), 139–162 (1987)
13. L. Caffarelli, A Harnack inequality approach to the regularity of free boundaries. II. Flat free boundaries are Lipschitz. Commun. Pure Appl. Math. **42**(1), 55–78 (1989)
14. L. Caffarelli, A Harnack inequality approach to the regularity of free boundaries. Part III: Existence theory, compactness, and dependence on X. Ann. Scuola Norm. Sup. Pisa **15**(4), 583–602 (1988)
15. L. Caffarelli, S. Salsa, A geometric approach to free boundary problems, in *Graduate Studies in Mathematics*, vol. 68 (American Mathematical Society, Providence, 2005)
16. L.A. Caffarelli, D.S. Jerison, C.E. Kenig, Global energy minimizers for free boundary problems and full regularity in three dimensions. Contemp. Math. **350**, 83–97 (2004)

© The Author(s) 2023
B. Velichkov, *Regularity of the One-phase Free Boundaries*,
Lecture Notes of the Unione Matematica Italiana 28,
https://doi.org/10.1007/978-3-031-13238-4

17. H. Chang-Lara, O. Savin, Boundary regularity for the free boundary in the one-phase problem. Comtemp. Math. (2017). Preprint. ArXiv:1709.03371

18. D. Danielli, A. Petrosyan, A minimum problem with free boundary for a degenerate quasilinear operator. Calc. Var. PDE **23**(1), 97–124 (2005)

19. G. David, T. Toro, Regularity for almost minimizers with free boundary. Calc. Var. PDE **54**(1), 455–524 (2015)

20. G. David, M. Engelstein, T. Toro, Free boundary regularity for almost-minimizers. Arxiv e-prints (2017)

21. G. De Philippis, B. Velichkov, Existence and regularity of minimizers for some spectral optimization problems with perimeter constraint. Appl. Math. Optim. **69**(2), 199–231 (2014)

22. G. De Philippis, J. Lamboley, M. Pierre, B. Velichkov, Regularity of minimizers of shape optimization problems involving perimeter. J. Math. Pure. Appl. **109**, 147–181 (2018)

23. D. De Silva, Free boundary regularity for a problem with right hand side. Interfaces Free Bound. **13** (2), 223–238 (2011)

24. D. De Silva, D. Jerison, A singular energy minimizing free boundary. J. Reine Angew. Math. **635**, 1–21 (2009)

25. D. De Silva, O. Savin, Almost minimizers of the one-phase free boundary problem. Commun. PDE **45**(8), 913–930 (2020)

26. D. De Silva, F. Ferrari, S. Salsa, Two-phase problems with distributed source: regularity of the free boundary. Anal. PDE **7**(2), 267–310 (2014)

27. N. Edelen, M. Engelstein, Quantitative stratification for some free boundary problems. Trans. Amer. Math. Soc. **371**, 2043–2072 (2019)

28. M. Engelstein, L. Spolaor, B. Velichkov, (Log-)epiperimetric inequality and regularity over smooth cones for almost area-minimizing currents. Geom. Topol. **3**, 513–540 (2019)

29. M. Engelstein, L. Spolaor, B. Velichkov, Uniqueness of the blow-up at isolated singularities for the Alt-Caffarelli functional. Duke Math. J. **169**(8), 1541–1601 (2020)

30. L.C. Evans, *Partial Differential Equations*, vol. 19 (American Mathematical Society, Providence, 1998)

31. L.C. Evans, R.F. Gariepy, Measure Theory and Fine Properties of Functions, 2nd revised edn. (CRC Press, Boca Raton, 2015)

32. H. Federer, The singular sets of area minimizing rectifiable currents with codimension one and of area minimizing flat chains modulo two with arbitrary codimension. Bull. Amer. Math. Soc. **76**, 767–771 (1970)

33. A. Friedman, Variational principles and free-boundary problems. *Dover Books on Mathematics* (John Wiley, New York, 1982)

34. N. Garofalo, F.-H. Lin, Monotonicity properties of variational integrals, A_p weights and unique continuation. Indiana Univ. Math. J. **35**, 245–268 (1986)

35. M. Giaquinta, E. Giusti, Global $C^{1,\alpha}$-regularity for second order quasilinear elliptic equations in divergence form. J. Reine Angew. Math. **351**, 55–65 (1984)

36. A. Henrot, M. Pierre, *Variation et Optimisation de Formes. Une Analyse Géométrique.* Mathématiques & Applications, vol. 48 (Springer, Berlin, 2005)

37. D.S. Jerison, O. Savin, Some remarks on stability of cones for the one phase free boundary problem. Geom. Funct. Anal. **25**, 1240–1257 (2015)

38. D. Kriventsov, F.H. Lin, Regularity for shape optimizers: the nondegenerate case. Commun. Pure Appl. Math. **71**(8), 1535–1596 (2018)

39. D. Kriventsov, F.H. Lin, Regularity for shape optimizers: the degenerate case. Commun. Pure Appl. Math. **72**(8), 1678–1721 (2019)

40. E. Lieb, M. Loss, *Analysis. Graduate Studies in Mathematics* (American Mathematical Society, Providence, 1997).

41. D. Mazzoleni, S. Terracini, B. Velichkov, Regularity of the optimal sets for some spectral functionals. Geom. Funct. Anal. **27**(2), 373–426 (2017)

42. D. Mazzoleni, S. Terracini, B. Velichkov, Regularity of the free boundary for the vectorial Bernoulli problem. Anal. PDE **13**(3), 741–764 (2020)

43. F. Maggi, *Sets of Finite Perimeter and Geometric Variational Problems: An Introduction to Geometric Measure Theory*, vol. 135 (Cambridge University Press, Cambridge, 2012)
44. A. Naber, D. Valtorta, Rectifiable-reifenberg and the regularity of stationary and minimizing harmonic maps. Ann. Math. **185**, 1–97 (2017)
45. E.R. Reifenberg, An epiperimetric inequality related to the analyticity of minimal surfaces. Ann. Math. **80**, 1–14 (1964)
46. E. Russ, B. Trey, B. Velichkov, Existence and regularity of optimal shapes for elliptic operator with drift. Calc. Var. PDE **58**(6), 199 (2019)
47. J. Serrin, A symmetry problem in potential theory. Arch. Rat. Mech. Anal. **43**, 304–318 (1971)
48. L. Simon, *Lectures on Geometric Measure Theory*. Proceedings of the Centre for Mathematical Analysis, vol. 3 (Australian National University, Canberra, 1983)
49. L. Spolaor, B. Velichkov, An epiperimetric inequality for the regularity of some free boundary problems: the 2-dimensional case. Commun. Pure Appl. Math. **72**(2), 375–421 (2019)
50. L. Spolaor, B. Trey, B. Velichkov, Free boundary regularity for a multiphase shape optimization problem. Commun. PDE **45**(2), 77–108 (2020)
51. B. Velichkov, *Existence and Regularity Results for Some Shape Optimization Problems*. Edizioni della Normale. Tesi, vol. 19 (Springer, Berlin, 2015). ISBN 978-88-7642-526-4
52. G.S. Weiss, Partial regularity for a minimum problem with free boundary. J. Geom. Anal. **9**(2), 317–326 (1999)
53. G.S. Weiss, A homogeneity improvement approach to the obstacle problem. Invent. Math. **138**(1), 23–50 (1999)

Index

© The Author(s) 2023
B. Velichkov, *Regularity of the One-phase Free Boundaries*,
Lecture Notes of the Unione Matematica Italiana 28,
https://doi.org/10.1007/978-3-031-13238-4

LECTURE NOTES OF THE UNIONE MATEMATICA ITALIANA

Editor in Chief: Ciro Ciliberto and Susanna Terracini

Editorial Policy

1. The UMI Lecture Notes aim to report new developments in all areas of mathematics and their applications - quickly, informally and at a high level. Mathematical texts analysing new developments in modelling and numerical simulation are also welcome.

2. Manuscripts should be submitted to
   Redazione Lecture Notes U.M.I.
   umi@dm.unibo.it
   and possibly to one of the editors of the Board informing, in this case, the Redazione about the submission. In general, manuscripts will be sent out to external referees for evaluation. If a decision cannot yet be reached on the basis of the first 2 reports, further referees may be contacted. The author will be informed of this. A final decision to publish can be made only on the basis of the complete manuscript, however a refereeing process leading to a preliminary decision can be based on a prefinal or incomplete manuscript. The strict minimum amount of material that will be considered should include a detailed outline describing the planned contents of each chapter, a bibliography and several sample chapters.

3. Manuscripts should in general be submitted in English. Final manuscripts should contain at least 100 pages of mathematical text and should always include

   – a table of contents;
   – an informative introduction, with adequate motivation and perhaps some historical remarks: it should be accessible to a
     reader not intimately familiar with the topic treated;
   – a subject index: as a rule this is genuinely helpful for the reader.

4. For evaluation purposes, please submit manuscripts in electronic form, preferably as pdf- or zipped ps-files. Authors are asked, if their manuscript is accepted for publication, to use the LaTeX2e style files available from Springer's web-server at
   ftp://ftp.springer.de/pub/tex/latex/svmonot1/ for monographs
   and at
   ftp://ftp.springer.de/pub/tex/latex/svmultt1/ for multi-authored volumes

5. Authors receive a total of 50 free copies of their volume, but no royalties. They are entitled to a discount of 33.3% on the price of Springer books purchased for their personal use, if ordering directly from Springer.

6. Commitment to publish is made by letter of intent rather than by signing a formal contract. Springer-Verlag secures the copyright for each volume. Authors are free to reuse material contained in their LNM volumes in later publications: A brief written (or e-mail) request for formal permission is sufficient.

Printed in the United States
by Baker & Taylor Publisher Services